T0185684

Lecture Notes in Physics

Volume 962

The Lecture Notes in Physics

The series Lecture Notes in Physics (LNP), founded in 1969, reports new developments in physics research and teaching-quickly and informally, but with a high quality and the explicit aim to summarize and communicate current knowledge in an accessible way. Books published in this series are conceived as bridging material between advanced graduate textbooks and the forefront of research and to serve three purposes:

- to be a compact and modern up-to-date source of reference on a well-defined topic
- to serve as an accessible introduction to the field to postgraduate students and nonspecialist researchers from related areas
- to be a source of advanced teaching material for specialized seminars, courses and schools

Both monographs and multi-author volumes will be considered for publication. Edited volumes should, however, consist of a very limited number of contributions only. Proceedings will not be considered for LNP.

Volumes published in LNP are disseminated both in print and in electronic formats, the electronic archive being available at springerlink.com. The series content is indexed, abstracted and referenced by many abstracting and information services, bibliographic networks, subscription agencies, library networks, and consortia.

Proposals should be sent to a member of the Editorial Board, or directly to the managing editor at Springer:

Lisa Scalone
Springer Nature
Physics Editorial Department
Tiergartenstrasse 17
69121 Heidelberg, Germany
Lisa.Scalone@springernature.com

More information about this series at http://www.springer.com/series/5304

André Melzer

Physics of Dusty Plasmas

An Introduction

 Springer

André Melzer
Institut für Physik
Universität Greifswald
Greifswald, Germany

ISSN 0075-8450   ISSN 1616-6361   (electronic)
Lecture Notes in Physics
ISBN 978-3-030-20259-0   ISBN 978-3-030-20260-6   (eBook)
https://doi.org/10.1007/978-3-030-20260-6

This Springer imprint is published by the registered company Springer Nature Switzerland AG.
The registered company address is: Gewerbestrasse 11, 6330 Cham, Switzerland

Preface

These Lecture Notes are intended as an introductory course into the fascinating field of dusty plasmas. On the one hand, dusty plasmas provide an independent branch of plasma physics with specific phenomena. Among those are crystallization processes, various forces, nonreciprocal dynamics, and many more. On the other hand, the study of dusty plasmas allows to view from a different angle on various aspects of general plasma physics like the problem of shielding or the mechanism of acoustic waves in plasmas. Thus, it provides new insights into accepted plasma physics concepts, thereby substantiating, revising, and strengthening its understanding.

It is the aim of these Lecture Notes to introduce the new approaches, novel phenomena, inherent peculiarities, and specific concepts of dusty plasmas and to spark the reader's interest in this still growing and interesting field of physics.

These Notes have emerged from the lectures held by the author at the University of Greifswald over the last years. They are intended to cover the physics of dusty plasmas on a graduate level. A background in general plasma physics on an introductory level is beneficial, but not required.

I am especially thankful to Birger Buttenschön, Michael Himpel, Carsten Killer, Harald Krüger, Ramana Ikkurthi, Yuriy Ivanov, Sebastian Käding, Tobias Miksch, Matthias Mulsow, Marian Puttscher, André Schella, Stefan Schütt, and Matthias Wolter for their proofreading, discussions, and suggestions. I am deeply indebted to Alexander Piel (Kiel University) for his helpful comments as well as for his stimulating discussions over so many years.

Greifswald, Germany
March 2019

André Melzer

Contents

1 Introduction 1
  References 4

2 Charging of Dust Particles 7
  2.1 Outline 7
  2.2 OML Charging Currents 8
      2.2.1 Derivation of the OML Currents 9
      2.2.2 Discussion of the OML Model 13
      2.2.3 Streaming Ions 14
      2.2.4 Collisions 14
  2.3 Other Charging Currents 16
  2.4 The Particle as a Floating Probe 17
      2.4.1 OML Currents 17
      2.4.2 The Capacitance Model 19
      2.4.3 Floating Potential with Streaming Ions 20
      2.4.4 Floating Potential with Collisions 21
      2.4.5 Floating Potential with Secondary Electron Emission 22
  2.5 Temporal Evolution of the Particle Charge 23
      2.5.1 Charging in the rf Sheath 24
      2.5.2 Stochastic Fluctuations 26
  2.6 Influence of Many Particles and Electron Depletion 27
  References 29

3 Forces and Trapping of Dust Particles 31
  3.1 Gravity 31
  3.2 Electric Field Force 31
      3.2.1 Shielding 32
      3.2.2 Polarization Forces 33
  3.3 Ion Drag Force 35
      3.3.1 Qualitative Discussion: Barnes Model 36
      3.3.2 Quantitative Discussion: Hutchinson/Khrapak Model 38

3.4	Neutral Drag Force	39
3.5	Thermophoresis	40
3.6	Laser Forces	41
3.7	Compilation of Forces, Dust Levitation and Trapping	42
3.8	Vertical Oscillations	45
3.8.1	Linear Resonance and Charge Measurement	46
3.8.2	Parametric Resonances	50
3.8.3	Nonlinear Resonances	51
3.8.4	Delayed Charging	54
3.9	Other Charge Measurements	55
References		56

4 Dust Particle Interaction 59

4.1	Interaction in the Horizontal Plane	59
4.2	Vertical Interaction	61
4.2.1	The Wakefield Potential (Wave Model)	62
4.2.2	The Ion Focus (Particle Model)	65
4.3	Non-Reciprocal Attraction	67
4.4	Measurement of the Attractive Force	69
4.5	Oscillatory Instability of the Vertical Alignment	72
4.5.1	Solution of the Coupled Oscillations	73
4.5.2	Oscillatory Instability	75
4.5.3	Conditions for the Instability	77
References		80

5 Plasma Crystallization and Phase Transitions 81

5.1	The One-Component Plasma (OCP)	81
5.2	Yukawa Systems	83
5.3	Coulomb Crystallization in Dusty Plasmas	84
5.4	Crystallization in Bounded Systems	86
5.5	Structural Information	87
5.6	3D Crystals	91
5.7	The Plasma Crystal	92
5.8	The Phase Transition of the Plasma Crystal: Experiment	94
5.9	The Phase Transition of the Plasma Crystal: Simulations and Theory	96
References		98

6 Waves in Weakly Coupled Dusty Plasmas 101

6.1	Dust-Acoustic Wave Dispersion	101
6.2	Discussion of the Dispersion Relation	105
6.3	Experiments on Dust-Acoustic Waves	106
6.4	Ion-Flow Driven Dust-Acoustic Waves	107
6.5	Dust Ion-Acoustic Wave	110
6.6	Other Wave Types	112
References		113

7 Waves in Strongly Coupled Dusty Plasmas 115
   7.1 Compressional Mode in 1D 115
   7.2 Dust Lattice Waves in 2D 119
   7.3 Compressional 1D Dust Lattice Waves: Experiment 122
   7.4 Compressional 2D Dust Lattice Waves: Experiment 125
   7.5 Shear 2D Dust Lattice Waves: Experiment 125
   7.6 Mach Cones 127
   7.7 Transverse (Out-of-Plane) Dust Lattice Waves 129
   7.8 Dispersion Relation from Thermal Particle Motion 131
   7.9 A Note on Shielding 133
   References 136

8 Finite Dust Clusters 139
   8.1 Formation of Finite Dust Clusters 139
   8.2 Structure of 1D Dust Clusters 140
   8.3 Structure of 2D Finite Dust Clusters 141
   8.4 Structure of 3D Finite Dust Clusters 144
   8.5 Normal Modes in Finite Clusters 146
   8.6 Modes from Thermal Particle Motion 150
   8.7 Stability 152
   8.8 Ion-Focus-Induced Phase Transitions in 2D Clusters 154
   8.9 Laser-Heated Phase Transitions in 3D Clusters 156
   References 158

9 Dusty Plasmas and Magnetic Fields 161
   9.1 Magnetization Parameters 161
   9.2 Moderate Axial Fields: Rotation 163
   9.3 Moderate Horizontal Fields: Sideways Motion 165
   9.4 Strong Axial Fields and Microparticles 166
   9.5 Strong Axial Fields and Nanoparticles 168
   9.6 Roundup 169
   References 169

10 Diagnostic Methods in Dusty Plasmas 171
   10.1 Light Scattering from Particles 171
   10.2 Image Analysis 173
   10.3 Particle Image Velocimetry 176
   10.4 Scanning Video Microscopy 176
   10.5 Digital Holography 179
   10.6 Stereoscopy 181
      10.6.1 Viewing Geometry 183
      10.6.2 Multiset Triangulation 184
      10.6.3 Shake-the-Box-Algorithm 187
   10.7 Imaging Diagnostics for Nanoparticles 190
      10.7.1 Dust Size Diagnostics 190
      10.7.2 Dust Density Diagnostics 193
   References 195

11   Particle Growth in Dusty Plasmas and Applications 199
      11.1   Particle Growth Mechanisms 199
             11.1.1   Cluster Formation 201
             11.1.2   Agglomeration/Coagulation Phase 203
             11.1.3   Accretion Phase.. 204
      11.2   Technological Impacts of Dusty Plasmas 205
             11.2.1   Surface Contamination by Dust........................... 205
             11.2.2   Dust in Fusion Devices 206
             11.2.3   Formation of New Materials 208
             11.2.4   Polymorphous Silicon Films 209
      References... 210

12   Astrophysical Dusty Plasmas 213
      12.1   Noctilucent Clouds and PMSE 214
      12.2   Dust Streams from Jupiter 216
      12.3   Dust Orbits at Saturn 218
      12.4   Spokes in Saturn's Rings.. 219
      12.5   Mach Cones in Saturn's Rings...................................... 222
      References... 223

13   Summary 225

Glossary 227

Index 231

Chapter 1
Introduction

Dusty (or colloidal or complex or fine-particle) plasmas are a fascinating field of plasma physics. Dusty plasmas enable us to study basic plasma properties on an "atomic" kinetic level and the allow to visualize collective plasma behavior, like oscillations and waves. Moreover, a vast number of novel phenomena are found in these systems. New features in dusty plasmas range from Coulomb crystallization to new types of forces and waves. In these lecture notes, a general introduction to this active and growing field will be given.

Dusty plasmas consist of (solid) particles in the size range between nanometers and tens of micrometers immersed in a gaseous plasma environment of electrons, ions and neutrals. The particles are charged by the inflow of plasma electrons and ions or by other means. Thus, the charged dust particles serve as an additional plasma species with separate time and spatial scales. Hence, the study of the plasma-particle and particle-particle interaction, their fundamental properties and their collective effects open up new and interesting visions on plasma physics.

Dusty plasmas share a number of physical concepts and similarities with non-neutral plasmas, like pure ion plasmas in Paul or Penning traps, as well as with colloidal suspensions, where charged plastic particles are immersed in an aqueous solution. In analogy to these systems, the terms "colloidal plasmas" or "complex plasmas" are used for dusty plasmas more or less as synonyms. Also, "fine-particle plasma" is common. Here, we will stay with the most customary term "dusty plasmas" for simplicity.

The field of dusty plasmas originated from the study of astrophysical situations where dusty plasmas are ubiquitous, like interstellar clouds, the rings of the Giant planets or comet tails. In his book [1], Verheest characterizes dusty plasmas by: "If the claim is made that more than 99% of the observable universe is in the plasma state then it could be jokingly asserted that the remainder is dust". For instance, Saturn's B-ring consists of micron and submicron dust particles (and larger boulders) in the plasma environment of Saturn. There, peculiar features have been observed by the Voyager spacecraft (see Fig. 1.1a): Radially extended structures

© Springer Nature Switzerland AG 2019

A. Melzer, *Physics of Dusty Plasmas*, Lecture Notes in Physics 962,
https://doi.org/10.1007/978-3-030-20260-6_1

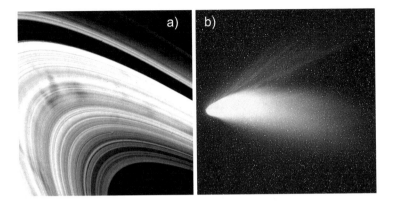

Fig. 1.1 (**a**) Spokes in Saturn's rings. The spokes are the radially extended dark features (Courtesy NASA/JPL-Caltech). (**b**) Plasma (blue) and dust (white) tail of comet Hale-Bopp. By E. Kolmhofer, H. Raab; Johannes-Kepler-Observatory, Linz, Austria (http://www.sternwarte.at) distributed under GFDL or CC BY-SA 4.0, from Wikimedia Commons

("spokes") develop within minutes and last for hours. This behavior cannot be explained by pure gravitational effects, but require the inclusion of particle-plasma interactions. Other examples of dusty astrophysical plasmas include comets and interstellar clouds. Comets are "dirty (dusty) snowballs" that evaporate under the influence of the solar wind. Comets usually form a plasma tail and the bright curved dust tail. The cometary dust is generally assumed to have a power-law size distribution with a high abundance of small grains (around 1 μm diameter) and a smaller number of larger particles [2]. The dust then interacts with the plasma tail near the comet head and the solar wind (see Fig. 1.1b). Finally, in interstellar clouds, typically dust particles with a size range of 0.01 to 10 μm are found. One of the interesting open questions, here, is how the charged dust particles influence star and planet formation.

A second origin of dusty plasmas leads to technological applications. For example, it is commonly said that in microchip manufacturing about 70% of the production steps require plasma processing techniques for etching or thin film deposition. Plasma processing allows to manufacture multi-layer devices and fine vertical structures that help to increase the packing density of integrated circuits. However, it was found at the end of the 1980s and beginning of the 1990s that, in reactive gases used for surface processing, dust particles can grow due to polymerization. The grown particles are trapped in the plasma near the wafers (see Fig. 1.2). It is obvious that dust particles falling on the etched structures ("killer particles") can lead to severe damages and malfunctions of the chips. Thus, the removal of dust particles from the discharge has been a crucial issue. Similarly, particles can also appear in fusion devices and their implication for operational performance of fusion machines is currently discussed. On the other hand, the incorporation of submicron dust particles in thin films leads to new materials, like

Fig. 1.2 Dust particles trapped above silicon wafers in a plasma processing discharge. Reprinted with permission from [3]. ©1991, American Vacuum Society

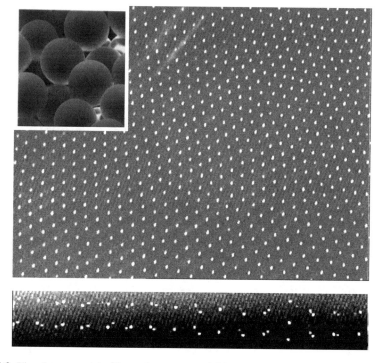

Fig. 1.3 The plasma crystal with two layers, top and side view. The inset shows a micrograph of the monodisperse plastic microspheres typically used in these experiments. From [7]

self-lubricating films with small embedded MoS_2 particles, or to polymorphous solar cells with increased efficiency.

In the 1990s, "plasma crystallization" and strong-coupling effects have been discovered in dusty plasmas [4–6] which has greatly promoted the studies of dusty plasma phenomena. There, it has been observed that dust particles can arrange in ordered crystal-like structures, the plasma crystal (see Fig. 1.3). In such laboratory experiments, monodisperse (i.e. particles of the same size) spherical particles of 1–10 μm diameter are trapped in gas discharges. The microspheres attain high negative charges of the order of 10^3 to 10^5 elementary charges due to the inflow of plasma electrons and ions. However, due to the extreme mass the charge-to-mass ratio of the dust particles is extremely small. Hence, the dynamic timescales are orders of magnitude slower than that of ions, not to mention electrons.

This makes dusty plasmas an ideal system to study particle behavior on the kinetic level of individual dust particles. The microspheres can be viewed individually due to their comparatively large size. The temporal and spatial scales for particle motion are ideal for studying the dynamics of dusty plasmas by video microscopy: interparticle distances are of the order of hundreds of microns, typical frequencies of the order of a few to a few ten Hertz. Moreover, due to the gaseous plasma environment the particle dynamics is only weakly damped.

Due to the high charges the electrostatic potential energy of the dust particles by far exceeds the thermal energy of the microspheres: the system is said to be strongly coupled. The strong-coupling regime is hardly reached in ordinary plasmas. Dusty plasmas therefore enable to study a vast variety of novel phenomena, like fluid and crystalline dust systems, phase transitions, strong-coupling effects, waves and Mach cones and many more. They provide a unique system bridging the fields of plasma physics, condensed matter and material science.

The above mentioned properties set the stage for these Lecture Notes on the physics of dusty plasmas. The topics presented in this introductory course are guided by the available experiments. The theoretical concepts are developed to a depth necessary to understand the experimental findings. An in-depth theoretical description is not intended, here. For further reading on other areas of interest the reader is referred to dedicated monographs [1, 8–11] or book chapters [12–15].

In these Lecture notes, I will first give an introduction to the fundamental properties of dusty plasmas which include particle charging, which is treated in Chap. 2, and forces on the dust particles in Chap. 3. The particle-particle interaction together with strongly coupled systems and phase transitions are described in Chaps. 4 and 5. Thereafter, collective effects like waves in weakly and strongly coupled systems will be discussed in Chaps. 6 and 7, respectively. The above concepts will then also be applied to systems consisting of only a small number of particles, so-called dust clusters, in Chap. 8. Magnetic field effects in dusty plasmas are presented in Chap. 9. Diagnostic techniques are illustrated in Chap. 10. Finally, in Chaps. 11 and 12 I will present phenomena in technological and astrophysical dusty plasmas that have already briefly been mentioned in this Introduction.

References

1. F. Verheest, *Waves in Dusty Space Plasmas* (Kluwer Academic Publishers, Dordrecht, 2000)
2. M. Fulle, *Comets II* (University of Arizona Press, Tucson, 2004), pp. 565–575
3. G.S. Selwyn, Plasma Sources Science and Technology **3**(3), 340 (1994). http://stacks.iop.org/0963-0252/3/i=3/a=016
4. J.H. Chu, I. Lin, Phys. Rev. Lett. **72**, 4009 (1994)

5. Y. Hayashi, K. Tachibana, Jpn. J. Appl. Phys. **33**, L804 (1994)
6. H. Thomas, G.E. Morfill, V. Demmel, J. Goree, B. Feuerbacher, D. Möhlmann, Phys. Rev. Lett. **73**, 652 (1994)
7. A. Melzer, A. Homann, A. Piel, Phys. Rev. E **53**, 2757 (1996)
8. A. Bouchoule (ed.), *Dusty Plasmas* (Wiley, Chichester, 1999)
9. P.K. Shukla, A.A. Mamun, *Introduction to Dusty Plasma Physics* (Institute of Physics Publishing, Bristol, 2002)
10. P. Bliokh, V. Sinitsin, V. Yaroshenko, *Dusty and Self-Gravitational Plasma in Space* (Kluwer Academic Publishers, Dordrecht, 1995)
11. A. Ivlev, H. Löwen, G. Morfill, C.P. Royall, *Complex Plasmas and Colloidal Dispersions: Particle-Resolved Studies of Classical Liquids and Solids*. Series in Soft Condensed Matter, vol. 5 (World Scientific, Singapore, 2012)
12. A. Piel, *Plasma Physics: An Introduction to Laboratory, Space, and Fusion Plasmas* (Springer, Heidelberg, 2010)
13. P.M. Bellan, *Fundamentals of Plasma Physics* (Cambridge University Press, Cambridge, 2006). https://doi.org/10.1017/CBO9780511807183
14. M. Bonitz, N. Horing, P. Ludwig, *Introduction to Complex Plasmas*. Springer Series on Atomic, Optical, and Plasma Physics (Springer, New York, 2010)
15. R. Hippler, H. Kersten, M. Schmidt, K.H. Schoenbach (eds.), *Low Temperature Plasmas* (Wiley-VCH, Weinheim, 2008)

Chapter 2
Charging of Dust Particles

The charge a particle attains in a plasma is certainly the most fundamental parameter of a dusty plasma. In the following, different charging mechanisms are discussed. Unlike for electrons or ions, the charge of a dust particle is not fixed, but determined by the ambient plasma parameters. This results in interesting phenomena in the temporal evolution of the dust charge. Further, important consequences for the dust charge in dense dust clouds arise.

2.1 Outline

The particle charging is, from the theoretical point of view, one of the most interesting but also difficult questions when all the details of the charging process of a dust particle are taken into account including trapping in the highly non-neutral, non-equilibrium sheath environment of a plasma with the presence of streaming ions and ion-neutral collisions. Hence, here, the charging model will be developed in steps starting from idealized cases to more complex situations.

The problem of particle charging is closely related to the theory of electrostatic (Langmuir) probes in plasmas. There, the task is to determine electron and ion densities, electron temperature etc. from the current-voltage characteristic of the probe. In contrast, for the problem of particle charging, we assume that these plasma parameters are known and the question is which potential on the dust grain is established by the currents onto the particle.

© Springer Nature Switzerland AG 2019
A. Melzer, *Physics of Dusty Plasmas*, Lecture Notes in Physics 962,
https://doi.org/10.1007/978-3-030-20260-6_2

In a plasma, the particle eventually attains a potential, the so-called floating potential ϕ_{fl} which is determined from the condition that at floating potential the sum of all currents to the particle vanishes, i.e.

$$\sum_{\ell} I_{\ell}(\phi_{fl}) = \frac{dQ_d}{dt} = 0, \qquad (2.1)$$

where I_{ℓ} denotes the different currents to the probe at the floating potential and Q_d is the dust charge. Since we are interested in the equilibrium charge (at least at the moment), the first task will be to determine this floating potential from these currents.

Currents to the particle may arise from the inflow of plasma electrons and ions or from the emission of secondary electrons and photoelectrons. Electron and ion collection will dominate in laboratory plasmas whereas secondary electron and photoelectron emission are considered important under astrophysical conditions. One can think of many other charging currents, like electron extraction due to strong electric fields etc., which however are considered unimportant in most cases and which will not be discussed here.

2.2 OML Charging Currents

We will start the discussion with electron and ion collection in the ideal case of an isotropic Maxwellian plasma environment at rest. The currents to the particle can be described by the OML ("orbital motion limit") model first derived by Mott-Smith and Langmuir in 1926 [1]. There, it is assumed that electrons and ions move towards the dust particle from infinity on collisionless orbits subject only to the electrostatic interaction with the dust particle. The OML problem is exactly equivalent to Kepler dynamics or Rutherford scattering.

Since the electrons are much more mobile than the ions the particle will generally charge negatively. In the calculation of the currents we will adopt that view and assume that the particle potential ϕ_p is negative with respect to the plasma potential, i.e. $\phi_p < 0$. Thus, the ions are the attracted and the electrons are the repelled species. The ion and electron OML currents are given by the following expressions (given here in advance for reference, they will be derived in the following section)

$$I_i = \pi a^2 n_i e \sqrt{\frac{8k_B T_i}{\pi m_i}} \left(1 - \frac{e\phi_p}{k_B T_i} \right)$$

$$I_e = -\pi a^2 n_e e \sqrt{\frac{8k_B T_e}{\pi m_e}} \exp\left(\frac{e\phi_p}{k_B T_e} \right), \qquad (2.2)$$

where a is the particle radius, e is the elementary charge, and k_B is Boltzmann's constant. Further, n_i, n_e are ion and electron densities, T_i, T_e the respective temperatures and m_i, m_e their masses. For the sake of completeness, the OML currents for positive particle potentials $\phi_p > 0$ read as

$$I_i = \pi a^2 n_i e \sqrt{\frac{8k_B T_i}{\pi m_i}} \exp\left(-\frac{e\phi_p}{k_B T_i}\right)$$

$$I_e = -\pi a^2 n_e e \sqrt{\frac{8k_B T_e}{\pi m_e}} \left(1 + \frac{e\phi_p}{k_B T_e}\right). \tag{2.3}$$

2.2.1 Derivation of the OML Currents

Ion Current As mentioned above, we will study here the usual case of $\phi_p < 0$ to derive the OML currents of Eq. (2.2). We start with the ion species that is attracted by the dust. Possible ion trajectories are depicted in Fig. 2.1. Ions with impact parameters smaller than a critical parameter b_c, i.e. $b < b_c$, will hit the particle, those with $b > b_c$ will only be deflected in the electric field of the dust. The critical parameter b_c is defined by the condition that the ion will hit the dust particle with grazing incidence.

At infinity the angular momentum of such an ion with the critical impact parameter b_c is

$$L = |\mathbf{r} \times \mathbf{p}| = m_i v_{i,0} b_c,$$

where $v_{i,0}$ is the ion velocity at infinity. Since this ion hits the dust at grazing incidence with velocity v_i its angular momentum at the particle surface is

$$L = m_i v_i a.$$

When there are no collisions the angular momentum of the ion is conserved.

The energy balance of the ions is the sum of kinetic and potential energy in the electrostatic potential of the dust particle, i.e.

$$\frac{1}{2} m_i v_{i,0}^2 = \frac{1}{2} m_i v_i^2 + e\phi_p.$$

Here, the energy of the ion at infinity and when arriving on the particle are balanced. Using the conservation of angular momentum the energy balance can be written as

$$\frac{1}{2} m_i v_{i,0}^2 = \frac{1}{2} m_i v_{i,0}^2 \left(\frac{v_i^2}{v_{i,0}^2} + \frac{e\phi_p}{(1/2)m_i v_{i,0}^2}\right) = \frac{1}{2} m_i v_{i,0}^2 \left(\frac{b_c^2}{a^2} + \frac{e\phi_p}{(1/2)m_i v_{i,0}^2}\right)$$

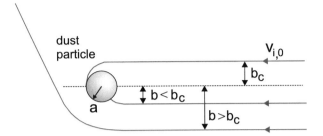

Fig. 2.1 Ion trajectories with different values of the impact parameter b

and thus

$$b_c^2 = a^2 \left(1 - \frac{2e\phi_p}{m_i v_{i,0}^2} \right). \tag{2.4}$$

We can then easily define the cross section for ion collection as

$$\sigma_c \equiv \pi b_c^2 = \pi a^2 \left(1 - \frac{2e\phi_p}{m_i v_{i,0}^2} \right). \tag{2.5}$$

It is seen that the cross section σ_c for ion collection is larger than the particle's geometric cross section $\sigma = \pi a^2$ (Remember: $\phi_p < 0$). Of course, this is due to the attraction by the dust as also easily seen from Fig. 2.1.

Now, the charging current to the particle by ions is given by

$$dI_i = \sigma_c(v_i) \, dj_i = \sigma_c(v_i) n_i e v_i f(v_i) dv_i,$$

where $j_i = n_i e v_i$ is the ion current density. The ion current has to be integrated over the ion velocity distribution function $f(v_i)$ that is assumed to be an isotropic Maxwellian in the following, i.e.

$$f(v_i) = 4\pi v_i^2 \left(\frac{m_i}{2\pi k_B T_i} \right)^{3/2} \exp \left(-\frac{\frac{1}{2} m_i v_i^2}{k_B T_i} \right) \quad \text{with} \quad \int_0^\infty f(v_i) dv_i = 1,$$

where T_i is the ion temperature. The charging current is then obtained by integration over the Maxwellian velocity distribution

$$I_i = 4\pi^2 a^2 n_i e \left(\frac{m_i}{2\pi k_B T_i} \right)^{3/2} \int_0^\infty \left(1 - \frac{2e\phi_p}{m_i v_i^2} \right) v_i^3 \exp \left(-\frac{\frac{1}{2} m_i v_i^2}{k_B T_i} \right) dv_i. \tag{2.6}$$

By rearranging we have

$$I_i = 4\pi^2 a^2 n_i e \left(\frac{m_i}{2\pi k_B T_i}\right)^{3/2} \times$$

$$\times \left[\underbrace{\int_0^\infty v_i^3 \exp\left(-\frac{\frac{1}{2}m_i v_i^2}{k_B T_i}\right) dv_i}_{F_1} - \underbrace{\int_0^\infty \frac{2e\phi_p}{m_i} v_i \exp\left(-\frac{\frac{1}{2}m_i v_i^2}{k_B T_i}\right) dv_i}_{F_2}\right]. \quad (2.7)$$

These two integrals can be easily evaluated resulting in[1]

$$F_1 = \frac{2(k_B T_i)^2}{m_i^2}$$

$$F_2 = \frac{2(k_B T_i)^2}{m_i^2} \frac{e\phi_p}{k_B T_i}.$$

Finally, collecting all terms the OML ion current is

$$I_i = \pi a^2 n_i e \sqrt{\frac{8k_B T_i}{\pi m_i}} \left(1 - \frac{e\phi_p}{k_B T_i}\right). \quad (2.8)$$

This relation can be interpreted as a product of the ion current density $j_i = n_i e v_{\text{th},i}$ at the ion thermal velocity[2]

$$v_{\text{th},i} = \sqrt{\frac{8k_B T_i}{\pi m_i}} \quad (2.9)$$

onto a dust particle and an effective cross section. The term $\pi a^2 (1 - e\phi_p/k_B T_i)$ describes this increased cross section of the dust particle at the ion thermal energy $k_B T_i$ (which is nearly the "real" ion thermal energy $(3/2)k_B T_i$).

---

[1]The definite integral is given by

$$\int_0^\infty x^n e^{-ax^2} dx = \frac{k!}{2a^{k+1}} \quad \text{for odd} \quad n = 2k+1.$$

[2]Throughout this book, the thermal velocity of all species is defined as $v_{\text{th}} = \sqrt{8k_B T/(\pi m)}$ with the respective temperature T and mass m.

Electron Current The electron current can be calculated in complete analogy to the ion's. The cross section of the dust particles for electrons is

$$b_c^2 = a^2 \left(1 + \frac{2e\phi_p}{m_e v_{e,0}^2} \right).$$

The difference to the ionic cross section is the "+"−sign in the bracket. This results in a reduction of the effective cross section compared to the dust particle area since the electrons are the repelled species (still $\phi_p < 0$). The electron current (with Maxwellian electrons) is then determined by

$$I_e = -4\pi^2 a^2 n_e e \left(\frac{m_e}{2\pi k_B T_e} \right)^{3/2} \int\limits_{v_{min}}^{\infty} \left(1 + \frac{2e\phi_p}{m_e v_e^2} \right) v_e^3 \exp\left(-\frac{\frac{1}{2}m_e v_e^2}{k_B T_e} \right) dv_e.$$

(2.10)

Here, the lower bound of the integral is $v_{min} = \sqrt{-2e\phi_p/m_e}$ instead of "0" in the ion current since only electrons that are energetic enough to overcome the electrostatic repulsion can reach the dust particle, i.e. electrons with $m_e v_e^2/2 > m_e v_{min}^2/2 = -e\phi_p$ are energetically able to arrive at the dust.

For the electron currents the integrals F_1 and F_2 analogous to Eq. (2.7) result in[3]

$$F_1 = \frac{2(k_B T_e)^2}{m_e^2} \exp\left(\frac{e\phi_p}{k_B T_e} \right) \left(1 - \frac{e\phi_p}{k_B T_e} \right)$$

$$F_2 = \frac{2(k_B T_e)^2}{m_e^2} \frac{e\phi_p}{k_B T_e} \exp\left(\frac{e\phi_p}{k_B T_e} \right)$$

The electron current finally is

$$I_e = -\pi a^2 n_e e \sqrt{\frac{8 k_B T_e}{\pi m_e}} \exp\left(\frac{e\phi_p}{k_B T_e} \right).$$

(2.11)

---

[3]The following integrals result in

$$\int\limits_{v}^{\infty} x e^{-ax^2} dx = \frac{1}{2a} e^{-av^2}$$

and

$$\int\limits_{v}^{\infty} x^3 e^{-ax^2} dx = \frac{1}{2a^2} \left(1 - av^2 \right) e^{-av^2}.$$

One could have obtained the same result for the electron current from a statistical description by arguing that the thermal electron flux onto the dust particle will be reduced by a Boltzmann factor $n_e \rightarrow n_e \exp(e\phi_p/k_B T_e)$ for the repelled electrons indicating that only the electrons of the high-energy tail of the Maxwellian distribution will reach the particle.

2.2.2 Discussion of the OML Model

General Our derivation was based on the assumption of **collisionless** ion trajectories. This is an essential ingredient in view of conservation of angular momentum. Further, the calculations were done for an **isotropic** situation where we have used the isotropic form of the Maxwellian velocity distribution function. Another assumption is that the distribution function for the electrons and ions is **Maxwellian** in the first place.

However, all of these conditions are often violated in plasma discharges. Typically, the ion mean-free path is often not much larger than the Debye length, especially in discharges in noble gases that have very large cross sections for ion-neutral charge exchange collisions. Moreover, the particles are trapped in the space charge sheath of a plasma discharge (as we will see below) where the ion motion is directed towards the electrodes. Finally, the distribution functions of electrons is in plasma discharges often better characterized by a bi-Maxwellian or Druyvesteyn-like distribution rather than a pure Maxwellian [2]. Hence, especially the high-energy tail of the electron distribution that determines the particle charge can strongly deviate from the Maxwellian assumption. For a full treatment of all these processes, thorough numerical simulations [3, 4] are required. However, the effect of streaming ions and ion-neutral collisions will be further discussed below using approximations to get a feeling for these effects.

Energy Barrier in the OML Model An interesting point is that for the OML motion an angular momentum barrier exists: Certain ions that from energy balance considerations should fall onto the particle will actually not reach the particle [5]. The reason is that the effective potential $U_{eff} = e\phi - L^2/2mr^2$ has a "hump" at distances $r > a$ where the ions are already reflected. Their distance of closest approach thus is not the particle surface. They do not come close to the region near the particle where the above discussed energy and angular momentum balance hold. This hump in the effective potential appears only for a small range of angular momenta and energies. For particles much smaller than the Debye length ($a \ll \lambda_D$) that fraction of reflected ions is negligible and the OML results are still valid [6, 7].

Radial Motion Theory In Langmuir probe theory a number of other models are available to describe the current to a (spherical) probe, like e.g. the radial motion theory (ABR) by Allen et al. [8] where the ions start from infinity with no kinetic energy and are accelerated radially towards the probe (or the dust particle). However, it has been shown [9] that in the ABR model the particle floating potential tends to zero for $a \ll \lambda_D$ which is unphysical. On the contrary, OML gives a finite particle potential independent of particle size (see Sect. 2.4.1).

2.2.3 Streaming Ions

In many cases, like in the sheath environment of a discharge, the ions have a drift velocity u_i that can be much larger than the thermal ion velocity $v_{th,i}$. Thus, the ion species is no longer described by an isotropic Maxwellian distribution. In that case, the ion current is modified. The ion current to a dust particle can then be written in the form

$$I_i = \pi a^2 n_i e v_{th,i} f(u_i) , \tag{2.12}$$

where $f(u_i)$ is a (rather complicated) function of the ion drift velocity [10], namely

$$f(x) = \sqrt{\frac{\pi}{4}} x \left[\left(1 + \frac{1}{2x^2} - \frac{e\phi_p}{k_B T_i x^2} \right) \operatorname{erf}(x) + \frac{1}{\sqrt{\pi}\, x} e^{-x^2} \right],$$

where $x = u_i / \sqrt{2 k_B T_i / m_i}$ and $\operatorname{erf}(x)$ is the error function.[4] For ions with high streaming velocity $u_i \gg v_{th,i}$ the ion current reduces to [10]

$$I_i = \pi a^2 n_i e u_i \left(1 - \frac{2e\phi}{m_i u_i^2} \right) , \tag{2.13}$$

which is obtained by replacing the thermal ion energy $k_B T_i$ in Eq. (2.2) by the kinetic energy of the drifting ions $m_i u_i^2 / 2$ and the thermal velocity $v_{th,i}$ by the drift velocity u_i. The effect of streaming ions is further discussed in Sect. 2.4.3.

2.2.4 Collisions

Ion-neutral collisions considerably modify the ion current onto the dust since an ion that might have missed the dust particle loses its kinetic energy in that collision and

---

[4]The error function is given by

$$\operatorname{erf}(x) = \frac{2}{\sqrt{\pi}} \int_0^x \exp(-y^2)\, dy.$$

is subsequently accelerated towards the dust. Further, the presence of collisions will lead to ions trapped in the electrostatic potential well performing Keplerian orbits around the dust [11]. In plasma discharges in noble gases ion-neutral collisions are predominantly charge-exchange collisions where an energetic ion transfers its charge to a formerly neutral thermal atom

$$A_{\text{fast}}^{+} + A_{\text{slow}} \longrightarrow A_{\text{fast}} + A_{\text{slow}}^{+}.$$

To quantify the contribution of these collisional ions to the current, one might estimate that every ion-neutral collision in a sphere of a certain radius R_0 around the dust leads to a collection of this ion. The probability of such a collision is roughly R_0/ℓ_{mfp} when independent collisions can be assumed. Here, $\ell_{\text{mfp}} = 1/(n_n\sigma_{\text{in}})$ is the mean free path for ion-neutral collisions, where n_n is the neutral gas density and σ_{in} is the ion-neutral (charge-exchange) collision cross section. The thermal current of ions through a sphere of radius R_0 is simply $I = \pi R_0^2 n_i e v_{\text{th,i}}$ yielding a collisional current of

$$I_{\text{coll}} = \pi R_0^2 n_i e v_{\text{th,i}} \frac{R_0}{\ell_{\text{mfp}}}.$$

Hence, the total ion current can be assumed [6] to be the sum of the ion collection current in Eq. (2.2) and the collision current

$$I_i = \pi a^2 n_i e v_{\text{th,i}} \left(1 - \frac{e\phi_p}{k_B T_i} + \frac{R_0^3}{a^2 \ell_{\text{mfp}}} \right). \tag{2.14}$$

A reasonable size R_0 would be where the ion-dust interaction energy is of the order of the ion thermal energy. Assuming a Debye-Hückel interaction of shielding length λ_D

$$\frac{k_B T_i}{e} = \frac{Z_d e}{4\pi \varepsilon_0 R_0} e^{-R_0/\lambda_D},$$

one finds, to a good accuracy [12] that $(R_0/\lambda_D)^3 = 0.1(a/\lambda_D)^2(e\phi_p/k_B T_i)^2$ (thereby making also use of the capacitance model in Sect. 2.4.2). Combining this all, the ion charging current including collisions can be written as [12]

$$I_i = \pi a^2 n_i e v_{\text{th,i}} \left(1 - \frac{e\phi_p}{k_B T_i} + 0.1 \left(\frac{e\phi_p}{k_B T_i} \right)^2 \frac{\lambda_D}{\ell_{\text{mfp}}} \right). \tag{2.15}$$

The effect of collisions on the particle potential will be further discussed in Sect. 2.4.4.

2.3 Other Charging Currents

Two other possible charging mechanisms are described here which are considered important mostly for astrophysical situations [13]. These are charging by UV radiation and secondary electron emission. Both processes lead to electron emission from the dust particle and can thus be treated as a positive current to the particle.

Other possible charging mechanisms, like thermal emission, field emission, sputtering, or ion-induced electron emission are neglected, here, since they are usually not important for dusty plasmas in the laboratory or in space.

Photoelectron Emission The absorption of UV radiation releases photoelectrons and hence constitutes a positive charging current. The magnitude of the current depends on the photoemission yield δ^{ph} which gives the amount of released electrons per incoming photon and is specific for the dust material ($\delta^{ph} \approx 1$ for metals and $\delta^{ph} \ll 1$ for dielectrics). In many cases the released photoelectrons can be described by a Maxwellian distribution with a temperature T_ν. The charging current by photoemission is then given by Goertz [13]

$$I_\nu = \delta^{ph} \pi a^2 e \Phi_\nu \qquad \phi_p \leq 0$$

$$I_\nu = \delta^{ph} \pi a^2 e \Phi_\nu \exp\left(-\frac{e\phi_p}{k_B T_\nu}\right) \qquad \phi_p > 0, \qquad (2.16)$$

where Φ_ν is the flux of photoelectrons. Photoemission by UV radiation might be important for dust grains in astrophysical situations near stars. Also, this effect is made responsible for dust layers floating above the surface of the Moon.

Secondary Electron Emission Secondary electron emission from the impact of energetic electrons might also be important under certain conditions in astrophysical situations. The secondary electron emission coefficient $\delta^S(E_e)$ describes the number of released electrons per incoming electron. It strongly depends on the energy of the impinging electron E_e and is described by the empirical formula [14]

$$\delta^S(E_e) = 7.4 \delta_m^S \frac{E_e}{E_m} \exp\left(-2\sqrt{\frac{E_e}{E_m}}\right)$$

that has the maximum yield of δ_m^S at $E_e = E_m$. Here, δ_m^S and E_m are material-dependent quantities that are in the range of $\delta_m^S = 0.5\ldots30$ and $E_m = 0.1\ldots2\,\mathrm{keV}$. Since the maximum yield δ_m^S can be much larger than one, many more electrons can be released for each incoming electron. This may even lead to positively charged dust grains.

Meyer-Vernet [15] has calculated the current from secondary electron emission: Integrating the secondary yield over a Maxwellian distribution of incoming electrons leads to the following expression of the charging current

$$I_S = 3.78_m^S \pi a^2 n_e \sqrt{\frac{8k_B T_e}{\pi m_e}} \exp\left(\frac{e\phi_p}{k_B T_e}\right) F_{5,0}\left(\frac{E_m}{4k_B T_e}\right) \tag{2.17}$$

for $\phi_p < 0$, and

$$I_S = 3.78_m^S \pi a^2 n_e \sqrt{\frac{8k_B T_e}{\pi m_e}} \left(1 + \frac{e\phi_p}{k_B T_e}\right) \exp\left(\frac{e\phi_p}{k_B T_e} - \frac{e\phi_p}{k_B T_s}\right) F_{5,B}\left(\frac{E_m}{4k_B T_e}\right) \tag{2.18}$$

for $\phi_p > 0$. Here, T_s is the temperature of the released secondary electrons, $B = \sqrt{4e\phi_p/E_m}$ and

$$F_{5,Y}(x) = x^2 \int_Y^\infty u^5 e^{-(xu^2+u)}\, du$$

is an integral that can be evaluated numerically.

2.4 The Particle as a Floating Probe

As mentioned above, the floating potential of the particle is determined from the condition that all currents to the particle vanish, see Eq. (2.1). The resulting potential is now discussed for different situations.

2.4.1 OML Currents

In laboratory discharges usually only the collection currents need to be considered. The OML electron and ion currents as a function of particle potential ϕ_p are shown in Fig. 2.2. The floating potential is then found from the intersection of the electron and ion currents.

Analytically, the floating potential ϕ_{fl} is obtained by equalizing the electron and ion currents resulting in

$$1 - \frac{e\phi_{fl}}{k_B T_i} = \sqrt{\frac{m_i T_e}{m_e T_i} \frac{n_e}{n_i}} \exp\left(\frac{e\phi_{fl}}{k_B T_e}\right). \tag{2.19}$$

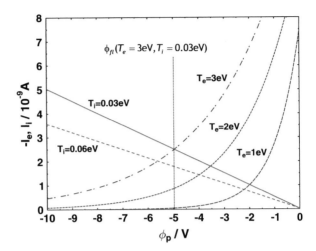

Fig. 2.2 OML currents onto a dust particle for different values of electron and ion temperatures. The intersection of electron and ion current gives the floating potential (the negative electron current is plotted as positive, here). Other parameters used here are $a = 4.7$ μm and $n_i = n_e = 1 \cdot 10^9$ cm^{-3}, which influence only the absolute magnitude of the currents, but not the floating potential. As an example, the floating potential $\phi_{fl} \approx -5$ V is indicated for the case of $T_e = 3$ eV and $T_i = 0.03$ eV

Table 2.1 Normalized floating potentials $\hat{\phi} = e\phi_{fl}/k_B T_e$ in quasineutral plasmas $n_e = n_i$ in hydrogen, helium and argon for different electron-to-ion temperature ratios

	$T_e/T_i =$						
	1	2	5	10	20	50	100
H	−2.504	−2.360	−2.114	−1.909	−1.700	−1.430	−1.236
He	−3.052	−2.885	−2.612	−2.388	−2.160	−1.862	−1.645
Ar	−3.994	−3.798	−3.491	−3.244	−2.992	−2.660	−2.414

This equation can be easily solved numerically for ϕ_{fl} for given values of the plasma parameters. In Table 2.1 the calculated floating potentials are shown for various discharge conditions. For the astrophysically important case of the isothermal ($T_e = T_i$) hydrogen plasma the well-known Spitzer value $\phi_{fl} = -2.5k_B T_e/e$ is approached, i.e. the normalized potential is $\hat{\phi} = e\phi_{fl}/k_B T_e = -2.5$. Under typical laboratory discharges in heavier gases with $T_e \gg T_i$ a good rule-of-thumb approximation is $\phi_{fl} \approx -(2\ldots2.5)k_B T_e/e$. However, for electron energy distributions with even a small suprathermal electron component the floating potential will be decisively different.

If the electron density is reduced with respect to the ion density $n_e < n_i$, e.g. in the sheath of a discharge, the electron charging current is reduced and the particle attains a more positive floating potential. This is shown in Fig. 2.3 where the floating potential is shown for a reduced electron density $\alpha = n_e/n_i$. For $n_e/n_i = 1$ the above value is retrieved (see Table 2.1), for $n_e/n_i \to 0$ even positively charged dust is found.

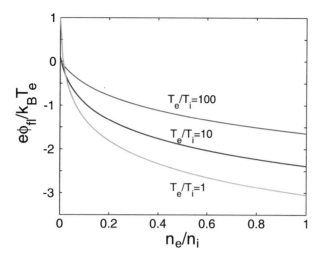

Fig. 2.3 Normalized floating potential of a dust particle $\hat{\phi} = e\phi_{\mathrm{fl}}/k_B T_e$ in helium for reduced electron density n_e/n_i

2.4.2   The Capacitance Model

After having determined the floating potential on the particle (at least in the case of pure OML currents) the charge on the particle has to be found. Therefore, the dust particle is considered as a spherical capacitor of capacitance C. The particle charge is then given by

$$Q_d = -Z_d e = C\phi_{\mathrm{fl}}, \tag{2.20}$$

where Z_d is the number of elementary charges on the dust. We will often treat Z_d as a positive number and deal with the fact that the dust (usually) is negatively charged by explicitly adding the minus sign.

In the vacuum case the capacitance of a sphere is

$$C = 4\pi\varepsilon_0 a.$$

In an ambient plasma with shielding length λ_D the capacitance is

$$C = 4\pi\varepsilon_0 a \left(1 + \frac{a}{\lambda_D}\right),$$

which for the typical case of $a \ll \lambda_D$ reduces to the vacuum value.

That means that the particle charge is

$$Z_d = \frac{4\pi \varepsilon_0 a}{e} \frac{k_B T_e}{e} \hat{\phi} \tag{2.21}$$

resulting in $Z_d = 695$ elementary charges per volt floating potential for a particle of $a = 1$ µm radius. With the above mentioned rule-of-thumb estimation of the floating potential $\phi_{fl} = -2k_B T_e/e$, i.e. $\hat{\phi} = -2$, an approximate formula of the dust charge is given by

$$Z_d = 1400 a_{\mu m} T_{e,eV} \tag{2.22}$$

with $T_{e,eV}$ being the electron temperature in electron volts and $a_{\mu m}$ the particle radius in microns.

2.4.3   Floating Potential with Streaming Ions

For streaming ions, the floating potential is obtained from the equality of the electron current and the ion current for drifting ions according to Eq. (2.12). For this case, the floating potential is shown in Fig. 2.4. One can see that for small drift velocities

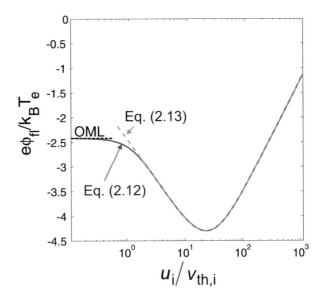

Fig. 2.4 Particle floating potential as a function of the streaming velocity of the ions. The floating potential has been normalized to the electron temperature, the drift velocity to the ion thermal velocity. The classical OML result is also indicated. The solid line is the full Eq. (2.12), the dashed line is the approximation of Eq. (2.13). Parameters: $T_e/T_i = 100$ in argon

up to the thermal velocity the floating potential is nearly unchanged and equal to the case without drift. Then, first a decrease of floating potential is observed (by about a factor of two) and after that a strong increase. Thus the particle first becomes more negative with respect to the OML case. The minimum floating potential is observed when the ions have approximately Bohm velocity $u_i \approx v_B = (k_B T_e / m_i)^{1/2}$. With further increased drift velocity the particles become less negatively charged. The first decrease of the potential is due to the reduction of the collection cross section $(1 - 2e\phi / m_i u_i^2)$ with increasing drift velocity. The increase of the potential at high drift velocities finally is due to the increased ion flux $n_i e u_i$. Moreover, it is seen that the approximated ion current according to Eq. (2.13) already works for relatively low drift velocities $u_i > v_{th,i}$.

2.4.4 Floating Potential with Collisions

In the case of collisions, the particle floating potential is derived by balancing the electron current and the collisional ion current according to Eq. (2.15). Figure 2.5 shows the floating potential of a dust particle in a collisional plasma background as a function of the mean free path ℓ_{mfp}. The OML result $e\phi_{fl} / k_B T_e = -2.414$ (compare Table 2.1) is shown as the dotted line. The red solid line indicates the

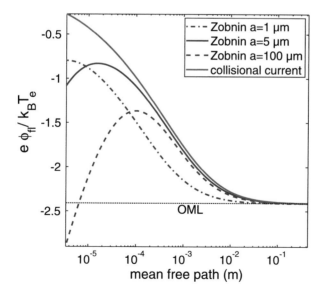

Fig. 2.5 Particle floating potential as a function of ion mean free path. Parameters: $n_i = n_e = 2 \cdot 10^8$ cm^{-3}, $T_e = 3$ eV, $T_i = 0.03$ eV in argon. The dotted line is the OML result. The solid red line indicates the result from the collisional ion current according to Eq. (2.15). The blue lines indicate the results of more sophisticated calculations by Zobnin et al. [16] for different particle sizes

influence of collisions using the collisional ion current of Eq. (2.15). For low collisionality (large mean free path) the OML result is retrieved. With shorter mean free path the particle potential becomes substantially less negative, the charge number substantially reduces. This is due to the collision-enhanced ion current to the particle.

For comparison, the results of a more sophisticated calculation and fitting procedure by Zobnin et al. [16] are also shown (see also [17]). They have revealed two effects: first, the floating potential depends on particle size (which is not found for OML or the collisional current). Second, at very small mean free path the particle potential again becomes more negative. This is due to the effect that there the ion motion towards the dust is effectively hampered by the very frequent ion-neutral collisions (hydrodynamic limit). However, the floating potential derived from our collisional ion current of Eq. (2.15) is in good agreement with the more sophisticated results down to mean free paths of the order of millimeter or even sub-millimeter for particle sizes in the micrometer range. Hence, collisions effectively reduce the particle floating potential and dust charge (up to a factor of 2) compared to the OML results.

2.4.5 Floating Potential with Secondary Electron Emission

If secondary electron emission is important, as for cosmic grains, the floating potential is determined from the condition

$$I_e + I_i + I_S = I_{tot} = 0.$$

Such a situation is shown in Fig. 2.6, where the secondary electron emission current I_S together with the OML charging currents I_e and I_i are shown as a function of the particle potential ϕ_p. Under the conditions chosen here the total current vanishes for three values of the floating potential. The floating potential thus is not a unique value, but is multivalued [15].

The two outer values of the floating potential ϕ_{p1} and ϕ_{p3} are stable roots, the middle ϕ_{p2} is unstable. The first stable root is always negative $\phi_{p1} < 0$, the second stable is always positive $\phi_{p3} > 0$. For these two outer roots

$$\left. \frac{d I_{tot}}{d\phi} \right|_{\phi_{p1,3}} < 0.$$

Thus, if a positive charge $+\Delta Q$, e.g. an ion, is added to the dust at the floating potential $\phi_{p1,3}$ the particle potential gets more positive $\phi_{p1,3} \rightarrow \phi_{p1,3} + \Delta\Phi_p$ which results in a negative total current to the particle that compensates the positive charge $+\Delta Q$. Therefore, $\phi_{p1,3}$ are stable roots. Correspondingly, ϕ_{p2} is unstable. Any fluctuation of the particle charge near ϕ_{p2} will switch the potential to either ϕ_{p1} or ϕ_{p3}.

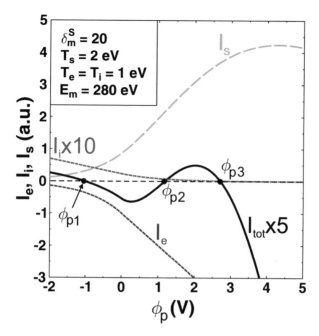

Fig. 2.6 Secondary electron emission current I_S, the OML charging currents I_e and I_i as well as the total current I_{tot} as a function of grain potential. The total current vanishes for three values of the floating potential, where the middle root is unstable, but the two extreme roots are stable. Hence, positive and negative grains can exist under the same plasma conditions. Parameters are as indicated in the inset. After [15]

That means that in the *same* plasma negatively as well as positively charged dust grains can exist [15]. The sign of the particle charge then depends on the charging history. Oppositely charged particles can then immediately coagulate in the plasma. This process might have an enormous influence on the growth mechanism of planetesimals in astrophysical dusty plasmas where secondary electron emission is present.

2.5   Temporal Evolution of the Particle Charge

So far, the particle charge has been considered as static. However, the temporal behavior of the dust charge can have a significant influence on the dynamic properties of the dust. The time scale for OML charging by ions τ_i is found from Eq. (2.1) by considering the ion charging current only

$$\frac{dQ_d}{dt} = I_i = \pi a^2 n_i e \sqrt{\frac{8 k_B T_i}{\pi m_i}} \left(1 - \frac{e\phi_p}{k_B T_i} \right).$$

Replacing the particle potential using the capacitor model this results in the differential equation

$$\frac{dQ_d}{dt} = \pi a^2 n_i e \sqrt{\frac{8k_B T_i}{\pi m_i}} \left(1 - \frac{Q_d e}{4\pi \varepsilon_0 a k_B T_i} \right). \tag{2.23}$$

This equation has the solution

$$Q_d(t) = Q_{d,0} e^{-t/\tau_i} + 4\pi \varepsilon_0 a \frac{k_B T_i}{e}$$

with the charging time scale

$$\tau_i = 4\pi \varepsilon_0 a \frac{k_B T_i}{e} \frac{1}{\pi a^2 e n_i v_{\mathrm{th},i}} = \frac{\varepsilon_0 \sqrt{2\pi m_i k_B T_i}}{e^2 a n_i}. \tag{2.24}$$

This time scale can be interpreted as the charging time $\tau = RC$ of the particle capacitance $C = 4\pi \varepsilon_0 a$ through the resistance $R = U/I$, where $U = k_B T_i/e$ is the typical electric potential of thermal ions and $I = \pi a^2 e n_e v_{\mathrm{th},i}$ is the typical thermal OML ion current to the particle. Or, alternatively from Eq. (2.8), $R = dU/dI$ at the floating potential. The electron charging time is given by the same formula where the ion quantities are replaced by those of the electrons:

$$\tau_e = 4\pi \varepsilon_0 a \frac{k_B T_e}{e} \frac{1}{\pi a^2 e n_e v_{\mathrm{th},e}} = \frac{\varepsilon_0 \sqrt{2\pi m_e k_B T_e}}{e^2 a n_e}. \tag{2.25}$$

Due to the higher electron mobility it follows that $\tau_e \ll \tau_i$, and the overall charging time is dominated by the slower ions. Typical charging times in laboratory discharges are of the order of microseconds (see also Figs. 2.7 and 2.8). This is still typically much faster than the dynamical time scale of the particles, so the particle charge should always be in equilibrium with the plasma conditions. However, we will discuss some situations in the following where the dust charging time scale plays a role.

2.5.1 Charging in the rf Sheath

In laboratory investigations on dusty plasmas the particles are usually trapped in the sheath of rf (radio-frequency) discharges, where the (time averaged) electric field force on the particle balances gravity (see Sect. 3.7). In the rf discharge, the electrons are able to follow the instantaneous rf voltage (which typically oscillates at 13.56 MHz), whereas the ions and the dust grains only react to the time averaged fields. In a simplified picture, the electron component of the plasma can be described as oscillating back and forth between the electrodes leading to a periodically

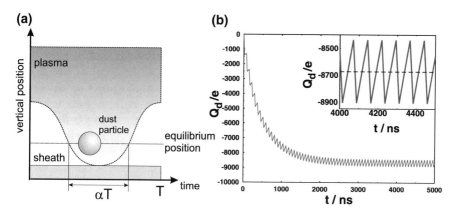

Fig. 2.7 Temporal evolution of a dust particle in the rf sheath. The inset is a magnification of the oscillating particle charge. Parameters: $a = 4.7$ μm, $kT_e = 3$ eV, $kT_i = 0.03$ eV, $n_i = 1 \times 10^9$ cm^{-3}, and the electron flooding is assumed as $\alpha = 25\%$ of the rf period. The equilibrium charge then is $Z_d = 8610$ negative elementary charges. After [18]

increasing and collapsing sheath at the electrodes. A dust particle trapped in the sheath therefore periodically "sees" a quasineutral plasma environment, when the sheath is flooded by electrons, and a pure ion sheath, when the sheath is expanded. Hence, the electron charging current to the dust particle will also be modulated by the rf frequency, thus

$$\dot{Q}_d = I_i + I_e \quad \text{for } 0 < t < \alpha T$$
$$\dot{Q}_d = I_i \qquad \text{for } \alpha T < t < T$$

where $\alpha = 0 \ldots 1$ is the fraction of the rf period $T = 75$ ns when the dust particle is in the "flooded" sheath (see Fig. 2.7a).

The temporal behavior of a dust particle in the sheath is shown in Fig. 2.7b for a particle immersed into the sheath starting from zero charge. In accordance with the charging time scale, see Eq. (2.24), the steady state charge is reached after a few microseconds. This is the same equilibrium charge as would be obtained from the OML model, Eq. (2.19), using $\alpha = n_e/n_i$.

Due to the periodic oscillation of the electron component, the dust particle charge is also modulated with the applied frequency of 13.56 MHz leading to a sawtooth behavior of the dust charge. In the inset one can see the steep electron charging towards more negative charge numbers and the slower ion discharging. After long times the particle charge is modulated by few hundred elementary charges at a mean particle charge of a few thousand. Hence, the charge modulation is a few percent at a high frequency. These modulations should not affect the particle dynamics on the time scale of the dust particle motion.

2.5.2 Stochastic Fluctuations

Besides the periodic modulation of the dust charge due to the rf sheath, the particle charge experiences stochastic fluctuations due to the discreteness of the charge carriers—electrons and ions. This effect has been investigated by Cui and Goree as well as Matsoukas and Russel [19, 20] simulating the collection of individual electrons or ions according to the OML currents. The probability of collecting an ion p_i or electron p_e is taken proportional to the ion (electron) current to the particle. The collection of electrons and ions is treated as a stochastic process. At the floating potential, obviously $p_i = p_e$. In Fig. 2.8 the temporal behavior of the dust charge due to stochastic charging processes starting from zero charge is shown.

In the beginning, at zero charge, the probability of collecting an electron is much, much higher than collecting an ion due to the higher mobility of the electrons. The particle gets negatively charged. After a certain time, the charging time according to Eq. (2.24), the probability of electron or ion collection becomes more and more equal to each other. After having reached the equilibrium value the dust charge fluctuates around this value with an amplitude of about 100 elementary charges. The equilibrium value Z_{eq} found in the stochastic approach is the same as for the static approach according to Eq. (2.19). It has been shown from simulations and analytical calculations that the charge fluctuations δZ_{rms} scale, as might be expected, as $\delta Z_{rms} \approx 0.5\sqrt{Z_{eq}}$. These fluctuations are small and quite high-frequent so that an influence on the particle dynamics is not presumed for micron sized particles. However, nanometer sized grains have equilibrium charges of the order of $Z_{eq} \approx 1$. Charge fluctuations can cause the particles to be charged even positively. Then one could also find positively and negatively charged dust in the same plasma which can lead to fast particle coagulation. This process might be of interest for particle growth in technological discharges (see Chap. 11).

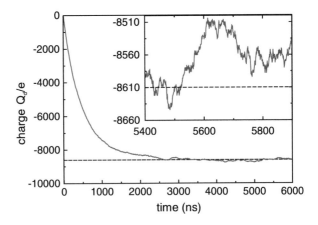

Fig. 2.8 Temporal behavior of the dust charge by stochastically collecting individual electrons and ions. The inset is a magnification of the dust charge fluctuations around equilibrium $Z_{eq} = 8610$. Parameters are the same as in Fig. 2.7. After [19]

2.6 Influence of Many Particles and Electron Depletion

So far, only the charging of single dust particles has been considered. Now, we would like to investigate the behavior of a dust cloud with many dust particles. When the dust density n_d is high, a considerable amount of the available electrons is bound to the dust. Thus, the density of free electrons in the plasma might be drastically reduced due to the presence of the dust which in turn influences the charging behavior of the dust [21].

For a quantitative description we consider a situation as described in Fig. 2.9a where a finite dust-containing plasma region is embedded in an infinite dust-free plasma. In the outer plasma, we find the undisturbed plasma density $n_{i,0} = n_{e,0}$, the dust density is zero ($n_d = 0$) and also the plasma potential ψ is defined as zero. In the dust-containing plasma cloud the plasma potential ψ differs from that in the outer plasma due to the presence of the charged dust particles (Fig. 2.9c). Now, the electron density in the dust cloud is assumed to adjust to the change in cloud potential with a Boltzmann factor, i.e.

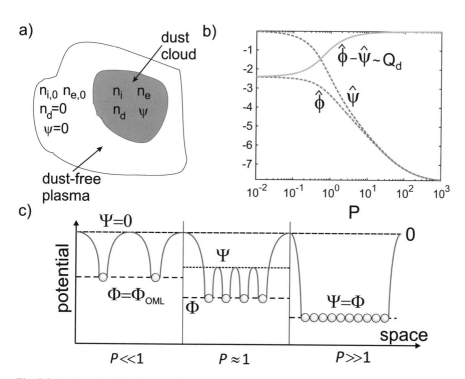

Fig. 2.9 (a) Sketch of the dust cloud in an extended pristine plasma. (b) Particle floating potential $\hat{\phi}$ and cloud potential $\hat{\psi}$ as a function of the Havnes parameter P. Parameters: $T_e/T_i = 100$ in argon. Note that the dust charge is proportional to $\hat{\phi} - \hat{\psi}$. (c) A one-dimensional section through the dense dust cloud. The particle potential ϕ and the cloud potential ψ in different regimes of the parameter P. After [22]

$$n_e = n_{e,0} \exp\left(\frac{e\psi}{k_B T_e}\right).$$

In contrast to the original discussion in [21] where also for the ions a Boltzmann distribution was assumed, for gas discharges it seems more reasonable to take the ion density as constant with $n_i = n_{i,0}$ [22].

The influence of the dust becomes noticeable when the quasineutrality condition

$$n_e e = n_i e - n_d Z_d e = n_i e - n_d C \phi_{fl} \qquad (2.26)$$

is affected by the dust. Here, one has to remember that the dust floating potential ϕ_{fl} has to be taken relative to the cloud potential. The positive ion charge density is balanced by either the free electrons or by the electrons bound on the dust particles. When the dust density is high then there are not enough electrons present to charge the particles to their single-particle negative floating potential. The particles thus are held at a more positive potential. This effect is called electron depletion, see Fig. 2.9c. Hence, the quasineutrality condition has to be coupled to the charging equation. Using the normalized dust floating potential $\hat{\phi}$ and plasma potential $\hat{\psi}$ with

$$\hat{\phi} = \frac{e\phi_{fl}}{k_B T_e} \quad \text{and} \quad \hat{\psi} = \frac{e\psi}{k_B T_e}$$

the quasineutrality condition and the charging equation are written as [22]

$$\exp\left(\hat{\psi}\right) - 1 - \underbrace{\frac{4\pi\varepsilon_0 a}{e} \frac{k_B T_e}{e} \frac{n_d}{n_{i,0}}}_{P}(\hat{\phi} - \hat{\psi}) = 0 \qquad (2.27)$$

$$\left(1 - \frac{T_e}{T_i}(\hat{\phi} - \hat{\psi})\right) - \sqrt{\frac{T_e m_i}{T_i m_e}} \exp\left(\hat{\phi}\right) = 0. \qquad (2.28)$$

The term containing $\exp(\hat{\psi})$ is the normalized electron density, and the particle potential is taken relative to the cloud potential as $\hat{\phi} - \hat{\psi}$. The Eqs. (2.27) and (2.28) have to be solved simultaneously for $\hat{\psi}$ and $\hat{\phi}$, i.e for the floating potential of the dust and the local plasma potential in the dust cloud. The dust influence is described in terms of the so-called Havnes parameter

$$P = \frac{4\pi\varepsilon_0 a}{e} \frac{k_B T_e}{e} \frac{n_d}{n_{i,0}} = 695\, T_{e,eV}\, a_{\mu m} \frac{n_d}{n_{i,0}}, \qquad (2.29)$$

where $T_{e,eV}$ is the electron temperature in electron volts and $a_{\mu m}$ is the particle radius in microns. By comparing with the dust charge, Eq. (2.21), the Havnes parameter can be approximately interpreted as the dust charge density in units of the undisturbed ion density

$$P \approx \frac{Z_d n_d}{n_{i,0}}.$$

Figure 2.9b illustrates the influence of the electron depletion with increasing Havnes parameter P. The single particle limit is obtained for $P \rightarrow 0$. Near $P \approx 1$ the floating potential on the particle and the cloud potential become more negative since the dust cloud as a whole does not benefit from an increased OML collection cross section [22, 23].

Much more important, however, is the evolution of the dust charge with the Havnes depletion parameter P. The dust charge is found from the relative particle potential $(\hat{\phi} - \hat{\psi})$ as, compare Eq. (2.21),

$$Z_d = \frac{4\pi \epsilon_0 a}{e} \frac{k_B T_e}{e} (\hat{\phi} - \hat{\psi}).$$

From Fig. 2.9b it is seen that the relative particle potential changes considerably to smaller (absolute) values, i.e. the dust becomes less and less (negatively) charged. The relative potential approaches $(\hat{\phi} - \hat{\psi}) \rightarrow 0$ for $P \gg 1$ reflecting the electron depletion: There are not enough electrons present as to charge all dust particles to the full single-particle potential, thus, the available electrons are distributed among the dust particles. Hence, in the limit $P \gg 1$, the free electron density $n_e \approx 0$ and consequently $Z_d \approx n_i/n_d$.

References

1. H.M. Mott-Smith, I. Langmuir, Phys. Rev. **28**, 727 (1926)
2. V.A. Godyak, R.B. Piejak, B.M. Alexandrovich, Plasma Sources Sci. Technol. **1**, 36 (1992)
3. V.R. Ikkurthi, K. Matyash, A. Melzer, R. Schneider, Phys. Plasmas **15**, 123704 (2008)
4. W.J. Miloch, S.V. Vladimirov, H.L. Pécseli, J. Trulsen, New J. Phys. **11**(4), 043005 (2009). http://stacks.iop.org/1367-2630/11/i=4/a=043005
5. J. Allen, B.M. Annaratone, U. deAngelis, J. Plasma Phys. **63**, 299 (2000)
6. M. Lampe, V. Gavrishchaka, G. Ganguli, G. Joyce, Phys. Rev. Lett. **86**, 5278 (2001)
7. C.T.N. Willis, M. Coppins, M. Bacharis, J.E. Allen, Plasma Sources Sci. Technol. **19**, 065022 (2010)
8. J. Allen, R. Boyd, P. Reynolds, Proc. Phys. Soc. **70**, 297 (1957)
9. R.V. Kennedy, J.E. Allen, J. Plasma Phys. **67**, 243 (2002)
10. E.C. Whipple, Rep. Prog. Phys. **44**, 1198 (1981)
11. J. Goree, Phys. Rev. Lett. **69**, 277 (1992)
12. S.A. Khrapak, S.V. Ratynskaia, A.V. Zobnin, A.D. Usachev, V.V. Yaroshenko, M.H. Thoma, M. Kretschmer, H. Höfner, G.E. Morfill, O.F. Petrov, V.E. Fortov, Phys. Rev. E **72**, 016406 (2005)
13. C.K. Goertz, Rev. Geophys. **27**, 271 (1989)
14. E.J. Sternglass, Sci. Pap. **1772** (1954)
15. N. Meyer-Vernet, Astron. Astrophys. **105**, 98 (1982)
16. A. Zobnin, A.D. Usachev, O.F. Petrov, V.E. Fortov, Phys. Plasmas **15**, 043705 (2008)

17. S. Khrapak, G. Morfill, Contrib. Plasma Phys. **49**, 148 (2009)
18. T. Nitter, T.K. Aslaksen, F. Melandsø, O. Havnes, IEEE Trans. Plasma Sci. **22**, 159 (1994)
19. C. Cui, J. Goree, IEEE Trans. Plasma Sci. **22**, 151 (1994)
20. T. Matsoukas, M. Russell, J. Appl. Phys. **77**, 4285 (1995)
21. O. Havnes, C.K. Goertz, G.E. Morfill, E. Grün, W. Ip, J. Geophys. Res. **92 A3**, 2281 (1987)
22. I. Goertz, F. Greiner, A. Piel, Phys. Plasmas **18**, 013703 (2011)
23. A. Piel, *Plasma Physics: An Introduction to Laboratory, Space, and Fusion Plasmas* (Springer, Heidelberg, 2010)

Chapter 3
Forces and Trapping of Dust Particles

In this chapter we will discuss the main forces acting on dust particles in a plasma discharge. These forces are gravity, electric field force, ion drag force, thermophoresis and neutral drag. After the analysis of the forces under plasma conditions the trapping of large (micron-sized) dust particles in the laboratory and under microgravity as well as of small (submicron) particles in plasma processing devices will be described. We are then in a position to discuss vertical oscillations in the plasma sheath and to derive a first set of methods for the charge measurement in the sheath of a plasma.

3.1 Gravity

The gravitational force simply is

$$\mathbf{F}_g = m_d \mathbf{g} = \frac{4}{3}\pi a^3 \varrho_d \mathbf{g}, \tag{3.1}$$

where $\mathbf{g}$ is the gravitational acceleration and ϱ_d is the mass density of the dust grains. Since this force scales with a^3 it is the dominant force for large particles in the micrometer range and becomes negligible for particles in the submicron and nanometer range.

3.2 Electric Field Force

Obviously, the electric field force due to an electric field $\mathbf{E}$

$$\mathbf{F}_E = Q_d \mathbf{E} = 4\pi \varepsilon_0 a \phi_{fl} \mathbf{E} \tag{3.2}$$

© Springer Nature Switzerland AG 2019
A. Melzer, *Physics of Dusty Plasmas*, Lecture Notes in Physics 962,
https://doi.org/10.1007/978-3-030-20260-6_3

is the governing force for charged particles. With the applied capacitor model this force scales linearly with the particle size. In the plasma sheath strong electric fields prevail that exert an electric field force that is large enough to levitate large micron-sized grains against gravity. In the plasma volume, due to quasineutrality, only small electric fields exist. Thus, in the plasma volume, particles can be trapped only when gravity is not important as for nanometer particles or for large particles under microgravity conditions.

3.2.1 Shielding

A subtlety should be discussed in more detail, here: Since in a plasma the (negative) dust particle is surrounded by a (positive) shielding cloud of ions, see Fig. 3.1a, one might think that the electric field force acts individually on the dust and on the shielding cloud in opposite directions, thereby canceling each other. In other words: the entire system of dust and shielding cloud is neutral and thus no net electric field force might exist, as depicted in Fig. 3.1b. If this reasoning was true there would be no action of the electric field force on the dust particle.

Hamaguchi and Farouki [1] have thoroughly discussed that point: To understand the behavior of the dust one has to carefully distinguish between the source and the effect of shielding. The shielding cloud is formed as a result of the presence of the electric field of the dust particle (and the external field). The shielding cloud reacts according to these fields. Hence, the shielding cloud is a response to the presence of the dust. When the dust moves from a place A in the plasma to a place B, the shielding cloud is not pushed from A to B by the dust. Rather, the ion cloud decays at A and re-forms at B. The shielding cloud is not "attached" to the dust by a force and thus there is no counterforce from the cloud onto the dust.

Consequently, the full force of Eq. (3.2) acts on the dust particle, the dust particle is moved by this force and the shielding cloud is formed according to the electric fields near the dust. The shielding cloud does not hinder the action of the electric

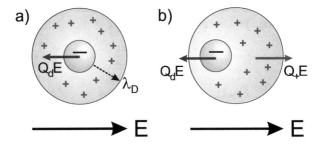

Fig. 3.1 Electric field force. (**a**) The electric field force acts on the dust particle even in presence of a shielding cloud. (**b**) "Wrong" picture of the dust particle and its shielding cloud under the influence of an external field E

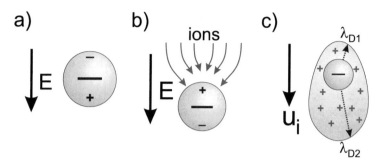

Fig. 3.2 Formation of a dipole moment on the dust particle (**a**) induced by an electric field or (**b**) by directed charging. (**c**) Polarization of the shielding cloud due to an ion streaming motion

field force on the dust. This reasoning holds for spherically symmetric shielding clouds. If the spherical symmetry of the shielding is distorted by the external field, additional polarization forces exist, that will be described now.

3.2.2   Polarization Forces

Polarization forces arise from dipole moments on the particles or the shielding cloud. In general, the polarization force on a dipole is

$$\mathbf{F}_{\text{dip}} = \nabla(\mathbf{p} \cdot \mathbf{E}), \tag{3.3}$$

where $\mathbf{p}$ is the dipole moment. Dipole moments on the particles can either be influenced by an external electric field or generated by directed charging processes (for dielectric particles) or a distortion of the shielding cloud.

The dipole moment on a dielectric particle induced by an external electric field $\mathbf{E}$, see Fig. 3.2a, is given by [2]

$$\mathbf{p} = 4\pi\varepsilon_0 a^3 \frac{\varepsilon - 1}{\varepsilon + 2}\mathbf{E},$$

where ε is the dielectric constant of the dust material. Due to the a^3 scaling the force due to an induced dipole becomes relevant only for extremely large particles.[1]

---

[1] As an estimate: Taking $\nabla E \sim E/\lambda_D$ with the Debye length λ_D and $E \sim k_B T_e/(e\lambda_D)$ yields $F_{\text{dip}} \sim F_E (a/\lambda_D)^2 \ll F_E$ for the typical situation $a \ll \lambda_D$.

Dipole moments may also arise from a directed charging (Fig. 3.2b) due to streaming ions where one side of the particle is hit more frequently by ions than the other side. For the case of a particle in the plasma sheath the top side is expected to be more positive than the bottom,[2] see also [3]. Note, that the dipole moments by directed charging and the induced dipole are opposite.

Hamaguchi and Farouki [1] have analyzed another situation which is important for the case of a plasma sheath. In the sheath there is a difference in electron and ion density which is sustained by an electric field $\mathbf{E}$. Consequently, the plasma densities can change over the size of the shielding cloud, thus a spatially dependent shielding length $\lambda_D(r)$ is considered in the model (see Fig. 3.2c). In the lower sheath of a discharge the electric field usually points downward from the positive plasma towards the negative electrodes. Near the plasma edge the electron and ion densities are larger than at the electrode. Thus, the shielding length on the top side λ_{D1} of a dust particle in the sheath is smaller than that on the bottom λ_{D2}.

The authors have solved the Debye-Hückel equation for the electrostatic potential $\phi(r)$ in this situation

$$\nabla^2\phi(r) - \frac{\phi(r)}{\lambda_D(r)^2} = \frac{Q_d}{\varepsilon_0}\delta(r)$$

and found the electric force on the dust particle including the contribution of the distorted shielding cloud as

$$\mathbf{F_E} = Q_d\mathbf{E} - \frac{Q_d^2}{8\pi\varepsilon_0}\frac{\nabla\lambda_D(r)}{\lambda_D^2}. \tag{3.4}$$

The force has two components. The first one is just the usual electric field force as in Eq. (3.2). The direction of the force is determined by the charge. The second term is due to the deformation of the shielding cloud. It is always in the direction of decreasing shielding length λ_D. The total force might thus be increased or decreased by the deformation. According to Hamaguchi and Farouki the shielding length to be used in Eq. (3.4) depends on the ion streaming velocity u_i. For subthermal ion drifts (i.e., $u_i < v_{th,i}$) ions can contribute to shielding and the relevant shielding length is the ion Debye length $\lambda_D = \lambda_{D,i}$. For supersonic ion streams ($u_i \gg v_{th,i}$) the ions are too fast to contribute to shielding and thus the relevant shielding length is the electron Debye length $\lambda_D = \lambda_{D,e}$ (see also Sects. 3.3.2 and 7.9). In most cases, however, the polarization forces are negligible, except for very large particles, see also [4].

---

[2]In this case, the dipole moment is of the order of $p \sim (4\pi\varepsilon_0 a)(k_B T_e/e)\,a$ [3], hence, $F_{dip} \sim F_E(a/\lambda_D)$.

3.3   Ion Drag Force

The next two forces to be discussed are drag forces which arise from a relative motion of a plasma species relative to the dust particle. In the case of the ion drag force there is an ion stream relative to the dust particle and the interaction is electrostatic, whereas for the neutral drag atoms or molecules of the neutral gas background drift past the dust and interact by direct collisions. The number of particles interacting with the dust per time interval dt is given by d$N = n\sigma v_{rel}$dt, where n is the density of the streaming species, σ is the cross section for interaction and v_{rel} is the relative velocity (see Fig. 3.3). Then, the drag force is given from the momentum transfer Δp per time interval as

$$\mathbf{F}_{\mathrm{drag}} = \frac{\mathrm{d}N}{\mathrm{d}t}\Delta\mathbf{p} = \Delta p\, n\sigma\mathbf{v}_{\mathrm{rel}}. \tag{3.5}$$

Now, we start with the ion drag force. The ion drag is due to ions streaming past the dust particle. An ion streaming motion can arise, e.g., from ambipolar diffusion or supersonic ion motion in the sheath. The ion drag consists of two parts, the collection force $\mathbf{F}_{\mathrm{dir}}$ due to ions directly hitting the dust and the Coulomb force $\mathbf{F}_{\mathrm{Coul}}$ due to Coulomb scattering of the ions in the electric field of the dust particle (see Fig. 2.1). The ion drag force will be discussed in two steps using, first, a simplified model for a qualitative understanding of the processes and, second, a more quantitative description is given. The ion drag force can have significant contributions in dusty plasmas. For example, the ion drag force is made responsible for the formation of the prominent "void" (see Sect. 3.7).

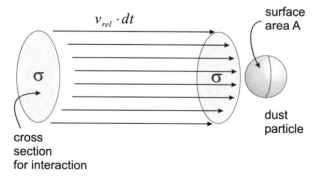

Fig. 3.3 Sketch indicating the number of plasma particles per unit time moving relative to the dust particle and interacting with it. Here, σ indicates the cross section for the relevant interaction of the plasma species with the dust

3.3.1 Qualitative Discussion: Barnes Model

In this first qualitative model, introduced by Barnes et al. [5], we note that the collection part of the ion drag force is just due to those ions which are also responsible for the ion contribution to charging of the dust and is thus given by

$$\mathbf{F}_{\text{dir}} = \pi a^2 m_i v_s n_i \mathbf{u}_i \left(1 - \frac{2e\phi_{\text{fl}}}{m_i v_s^2} \right). \tag{3.6}$$

Here, $m_i v_s$ is the momentum transfer of the ion at the "mean" velocity $v_s = (u_i^2 + v_{\text{th},i}^2)^{1/2}$ given by the ion thermal velocity $v_{\text{th},i}$ and the ion drift velocity u_i. The interaction cross section is that for ion collection defined in Eq. (2.5).

The Coulomb force is exerted by those ions which are not collected by the dust, but are deflected in the electric field of the dust grain. From plasma physics textbooks it is known that

$$\sigma = 4\pi b_{\pi/2}^2 \ln \Lambda = 4\pi b_{\pi/2}^2 \ln \frac{\lambda_D}{b_{\pi/2}}$$

is the cross section for ion-electron Coulomb collisions, where $b_{\pi/2} = e^2/(4\pi\varepsilon_0 m_i v^2)$ is the impact parameter for $90°$ deflection and $\ln \Lambda$ is the Coulomb logarithm. The Coulomb logarithm is due to truncation of the intrinsically infinite Coulomb collision cross section at the Debye length λ_D, i.e. Coulomb collisions outside the screening cloud are regarded unimportant. A cut-off towards small impact parameters is not necessary since electrons and ions are considered point-like.

For collisions with finite-sized dust particles the above Coulomb cross section has to be modified. The minimum collision parameter is b_c since ions with $b < b_c$ are absorbed by the particle and contribute to the direct force. Integrating the Coulomb collision cross section from b_c to λ_D yields [5]

$$\sigma = 4\pi b_{\pi/2}^2 \ln \left(\frac{\lambda_D^2 + b_{\pi/2}^2}{b_c^2 + b_{\pi/2}^2} \right)^{1/2}.$$

with the slightly modified $90°$ collision parameter

$$b_{\pi/2} = \frac{Z_d e^2}{4\pi\varepsilon_0 m_i v_s^2} = \frac{ae\phi_p}{m_i v_s^2}.$$

In the last equation we have used the capacitor model to replace the dust charge Z_d.

Thus, finally, the Coulomb force on a dust particle from streaming ions is

$$\mathbf{F}_{\text{Coul}} = 4\pi b_{\pi/2}^2 m_i n_i v_s \mathbf{u}_i \ln \left(\frac{\lambda_D^2 + b_{\pi/2}^2}{b_c^2 + b_{\pi/2}^2} \right)^{1/2} \tag{3.7}$$

$$= 2\pi \frac{a^2 e^2 \phi_p^2}{m_i v_s^3} n_i \mathbf{u}_i \ln \left(\frac{\lambda_D^2 + b_{\pi/2}^2}{b_c^2 + b_{\pi/2}^2} \right)$$

In this force calculation, still only ion trajectories within one Debye length λ_D around the dust are considered. Barnes et al. [5] leave open the question which Debye length has to be used, the electron or the ion Debye length. As a crude estimate, Eq. (3.7) with the electron Debye length $\lambda_{D,e}$ can be used [6].

Finally, the total ion drag force is just the sum of the direct and the Coulomb collision force

$$\mathbf{F}_{\text{ion}} = \mathbf{F}_{\text{dir}} + \mathbf{F}_{\text{Coul}} \tag{3.8}$$

and is directed along the ion streaming motion. The ion drag force is shown in Fig. 3.4. The ion drag shows a pronounced S-shape. For small drift velocities the force first increases, then decreases and finally increases again. The first maximum is obtained when the drift velocity is of the order of the thermal velocity of the ions. For small drift velocities only the Coulomb force needs to be considered. The increase of the force at very low velocities is due to the increase of the Coulomb logarithm (due to the decrease of $b_{\pi/2}$) with velocity. After the maximum the Coulomb force decreases approximately as $1/u_i^2$. For $u_i \gg v_{th,i}$ the direct, collection force becomes dominant and increases as u_i^2.

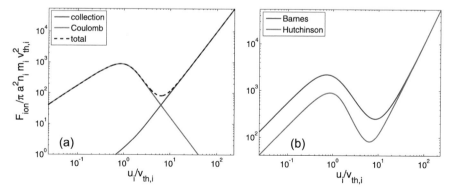

Fig. 3.4 Calculated ion drag force on a double-logarithmic scale. In (**a**) the total force and contribution of the ion collection and the Coulomb scattering are shown using the Hutchinson/Khrapak model. (**b**) Comparison of the qualitative Barnes model (using $\lambda_{D,e}$) with the sophisticated Hutchinson/Khrapak model

3.3.2 Quantitative Discussion: Hutchinson/Khrapak Model

There are, however, some difficulties associated with the Barnes model. First, for the highly charged dust grains also Coulomb collisions outside the Debye sphere have to be included [7] since the ions feel the electrostatic potential of the highly negative dust particle far further than just one Debye length, hence the Coulomb logarithm has to be modified. Moreover, as already became obvious in the above discussion, a second problem is associated with the correct Debye length at the different ion streaming velocities. Finally, especially at low ion streaming velocities one has to consider that the ion velocity is a combination of the ion thermal velocity and the drift velocity which is adequately described by a shifted Maxwellian distribution function.

The collection force with a shifted Maxwellian ion distribution is then written as (see e.g. [8, 9])

$$F_{\text{dir}} = \pi a^2 m_i n_i v_{\text{th,i}}^2 H(u), \tag{3.9}$$

where $H(u)$ which is a function of the normalized ion drift velocity $u = u_i/\sqrt{2kT_i/m_i}$ and the normalized particle potential $\chi = -e\phi_p/kT_i$ is given by[3]

$$H(u) = \frac{\sqrt{\pi}}{8u^2} \left[u \left(2u^2 + 1 + 2\chi \right) e^{-u^2} + \left(4u^4 + 4u^2 - 1 - 2(1 - 2u^2)\chi \right) \frac{\sqrt{\pi}}{2} \text{erf}(u) \right].$$

The function scales as $H(u) \to (\pi/4)u^2 = u_i^2/v_{\text{th,i}}^2$ for $u \gg 1$, hence $F_{\text{dir}} = \pi a^2 m_i n_i u_i^2$ which corresponds to the Barnes expression in Eq. (3.6) at high drift velocities. Hence, as shown in Fig. 3.4b, for $u_i \gg v_{\text{th,i}}$ the two models agree.

Similarly, the Coulomb force is now given by

$$F_{\text{Coul}} = 32 \frac{a^2 e^2 \phi_p^2}{m_i v_{\text{th,i}}^2} n_i \, G(u) \, \ln \Lambda \tag{3.10}$$

with the Chandrasekhar function $G(u) = [\text{erf}(u) - (2/\sqrt{3})ue^{-u^2}]/(2u^2)$. For strong ion drifts $u \gg 1$ the Chandrasekhar function scales as

$$G(u) \to \frac{1}{2u^2} = \frac{\pi}{16} \frac{v_{\text{th,i}}^2}{u_i^2}$$

yielding the same functional form as the Barnes expression in Eq. (3.7).

However, the strongest difference to the qualitative discussion above lies in the determination of the Coulomb logarithm $\ln \Lambda$. It has been shown by Khrapak et al.

---

[3]The different prefactor given here compared that in Ref. [9] is due to the different normalizations of the thermal velocity.

[8] and Hutchinson [9] that the Coulomb logarithm in the modified form

$$\ln \Lambda = \ln \frac{b_{\pi/2} + \lambda_s}{b_{\pi/2} + a} \qquad (3.11)$$

with a velocity dependent screening length already yields quite accurate results. The screening length

$$\lambda_s^2 = \frac{\lambda_{D,e}^2}{1 + 2kT_e/(m_i v_s^2)} + a^2 \qquad (3.12)$$

interpolates between the linearized Debye length at thermal ion velocities and the electron Debye length at high ion streaming velocities. This is reasonable since at higher streaming velocities the ions cannot participate in shielding (see Sect. 7.9 and also Sect. 3.2.2).

Extensive numerical calculations by Hutchinson [9] have revealed that for evaluating $b_{\pi/2} = ae\phi_p/(m_i v_{eff}^2)$ in the Coulomb logarithm of Eq. (3.11) and for the velocity dependent screening length λ_s in Eq. (3.12) an effective ion velocity of the form

$$v_{eff}^2 = 2\frac{kT_i}{m_i} + u_i^2 \left(1 + \left[\frac{u_i/\sqrt{kT_e/m_i}}{0.6 + 0.05\ln\mu + (\lambda_{D,e}/5a)(\sqrt{T_i/T_e} - 0.1)}\right]^3\right)$$

should be used. Here, μ is the atomic mass number of the ion.

The Hutchinson/Khrapak model yields quantitative results for the ion drag force. The total ion drag and its collection and Coulomb contribution using this model is shown in Fig. 3.4b. The comparison of the Hutchinson/Khrapak and the Barnes model indicates substantial differences of the order of a factor of 2 and more, especially for the slow ion drift velocities that are relevant in many experiments.

As complicated as the above discussions and equations already seem, they still refer to the case of collisionless ion trajectories. Simulations of the ion drag including ion-neutral collisions indicate a certain influence of collisions on the ion drag force [10–12]. Collisions lead to a slight increase (less than a factor of 2) of the ion drag force for moderate collisionalities [12].

3.4 Neutral Drag Force

The neutral drag is a friction force due to neutral gas atoms impinging on the dust particle. Neutral gas atoms or molecules that hit the dust lead to a slowing of the dust particle motion (when the dust is moving). Using Eq. (3.5) the force on a moving dust grain with velocity $\mathbf{v}_d$ is readily given as [13]

$$\mathbf{F}_n = -\delta\frac{4}{3}\pi a^2 m_n v_{th,n} n_n \mathbf{v}_d, \qquad (3.13)$$

where m_n, n_n and $v_{th,n}$ are the mass, the density and the thermal velocity of the neutral gas atoms, respectively. The cross section for interaction is just the geometrical particle cross section πa^2. The momentum transfer of the neutral gas onto the dust is of the order of $m_n v_{th,n}$ and slightly depends on how the gas atoms are reflected from the particle surface, e.g. diffuse, by specular reflection, isotropic etc. These differences are accounted for by the parameter δ which lies in the range between 1 (for specular reflection) and 1.44 (for diffuse reflection). Experiments [14, 15] on plastic microspheres in a plasma favor a value of $\delta = 1.44$.

This neutral drag force was derived by Epstein in 1924 for the analysis of the friction force in Millikan's famous oil drop experiment. We will often use that neutral drag in the form of a friction force, i.e.

$$\mathbf{F}_n = -m_d \beta \mathbf{v}_d \tag{3.14}$$

with

$$\beta = \delta \frac{8}{\pi} \frac{p}{a \rho_d v_{th,n}}. \tag{3.15}$$

Here, β is the (Epstein) friction coefficient and linearly depends on the gas pressure p. The friction coefficient is inversely proportional to the particle radius a which means that, in relation to their mass, smaller particles experience stronger damping than larger particles.[4]

3.5   Thermophoresis

The thermophoretic force acts on a dust particle due to a temperature gradient in the neutral gas. In an (over)simplified picture, it can be argued that neutral gas atoms from the "hotter" side of the gas hitting the dust grain transfer a larger momentum and thus exert a stronger force than atoms from the "colder" side of the gas. This leads to a force towards colder gas regions. However, the complete picture is more difficult and will not be discussed here. Following a rigorous treatment from gas kinetic theory [16], the thermophoretic force is found to be

$$\mathbf{F}_{th} = -\frac{32}{15} \frac{a^2 k_n}{v_{th,n}} \nabla T_n \tag{3.16}$$

---

[4]When defining the thermal velocity as $\tilde{v}_{th,n} = \sqrt{k_B T_n / m_n}$, then $\beta = \delta \sqrt{8/\pi}\, p/(a \rho_d \tilde{v}_{th,n})$.

with ∇T_n being the temperature gradient of the neutral gas and k_n the thermal conductivity of the gas.[5]

The thermophoretic force is considered to be important for sub-micron particles due to heating of the gas by the plasma discharge. It has been intentionally applied for the levitation of micron-sized particles using strong temperature gradients [17, 18] and for the formation of 3D Yukawa balls (see Sect. 8.4 and [19]).

3.6   Laser Forces

In dusty plasmas focused laser beams are used to manipulate dust particles. The origin of the laser-particle manipulation goes back to the pioneering works of Ashkin [20] who has confined particles in optical traps. These works also have led to the techniques of laser-cooling of ions which is needed to form, e.g. Bose-Einstein condensates.

The laser beam interacts with the dust particle and pushes the particle in the direction of the beam. The origin of this laser force in dusty plasmas can be due to two sources, the radiation pressure or the so-called photophoretic force.

The force exerted by radiation pressure is just the momentum transfer of the laser photons p_{ph} that hit the dust particle. The radiation pressure can be written as [20]

$$P_{rad} = \frac{1}{A}\frac{dp_{ph}}{dt} = \frac{1}{A}\frac{dN_{ph}}{dt}\frac{h}{\lambda} = \frac{1}{A}\frac{dN_{ph}}{dt}\frac{h\nu}{c} = \frac{I_L}{c}. \tag{3.17}$$

Here, N_{ph} is the number of photons hitting the dust particle and λ and ν are the laser wavelength and frequency, respectively. The intensity I_L of the laser beam is just the number of photons N_{ph} of energy $h\nu$ per time interval focused onto the cross section A.

The force exerted by the radiation pressure on the dust then is

$$F_{rad} = \gamma\frac{I_L}{c}A_d \tag{3.18}$$

where $A_d = \pi a^2$ is the geometric cross section of the dust particle and γ is a coefficient that takes the kind of interaction of the photons with the particle into account, i.e. $\gamma = 2$ for total reflection of the photons or $\gamma = 1$ for pure absorption.

For (partly) transparent particles, the radiation force also has a component perpendicular to the laser beam that results from the intensity profile of the beam (see Fig. 3.5). Typically, a laser beam has a profile that has maximum intensity in the center and then gradually decreases in the transverse direction. For a dust particle

---

[5]When defining the thermal velocity as $\tilde{v}_{th,n} = \sqrt{k_B T_n/m_n}$, then $F_{th} = -(16/15)\sqrt{\pi/2}$ $a^2 k_n/\tilde{v}_{th,n}\nabla T_n$.

Fig. 3.5 Scheme of the laser-particle interaction due to radiation pressure

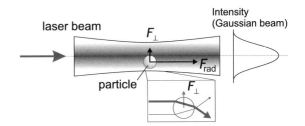

in the laser beam, this means that more photons go through the side of the particle that is closer to the center of the beam (the upper side of the particle in Fig. 3.5) compared to the other side (the lower side). The photons are deflected by the particle due to refraction. Since more photons on the upper side of the particles are deflected downwards than vice versa, an upward force (towards the center) arises. This means, the particle is "sucked" into the beam in the direction perpendicular to the beam. This of course helps in the application of laser manipulation in dusty plasmas. This "trapping" of particles in the beam has been applied in the optical trap experiments of Ashkin [20] and has also been observed in dusty plasma experiments.

The second mechanism, the photophoretic force, that yields a force from laser beam interaction is much more difficult to analyze quantitatively. There it is assumed that the laser heats the illuminated particle. Similar to the thermophoretic force discussed above, neutral particles that impinge on the "hot" side of the dust are reflected at higher velocities than on the cold side. This leads to a force away from the "hot" side of the particle. However, the temperature distribution across the particle is very difficult to determine.

Irrespective of the exact mechanism, lasers have been successfully applied in dusty plasmas to drive and manipulate particles (see e.g. [14, 21–23]) and in many cases, the radiation pressure force, Eq. (3.18) was sufficient to explain the particle motion.

3.7 Compilation of Forces, Dust Levitation and Trapping

After the description of the relevant forces on the dust particles, now the question is how these forces can lead to particle levitation and trapping. Therefore, the strength of these forces as a function of particle size is calculated under the typical experimental conditions of a low-power rf Ar discharge (Fig. 3.6). The neutral drag is of interest only for moving particles and not for the identification of stable equilibria. Thus, the neutral drag is not considered, here.

For micron-sized dust particles the dominant forces are gravity and electric field force. The thermophoretic force is already quite large when assuming a moderate

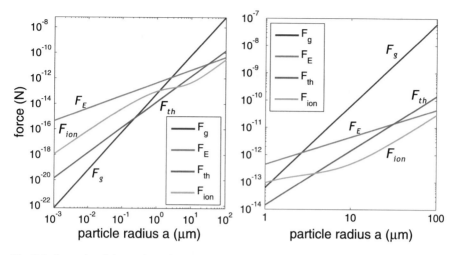

Fig. 3.6 Strength of the various forces as a function of dust particle radius. The employed parameters are: $\varrho_d = 1500$ kg/m^3, $T_e = 2$ eV, $\phi_{fl} = -4$ V, $E = 1000$ V/m, $n_i = 10^{15}$ m^{-3}, $u_i = v_{th,i} = v_{th,n} = 400$ m/s, $\nabla T_n = 200$ K/m, $k_n = 0.016$ kg m s^{-3} (Ar). The right panel is the same as the left panel, only zoomed to a restricted particle size

temperature gradient of 200 K/m.[6] The ion drag force is smaller than the electric field force and gravity under the chosen conditions. Hence, for dust particles above 1 micron in diameter and with small temperature gradients electric field force and gravity are most important. (Please note, that the electric field is considered as constant, here. In a discharge, the dust will move to a position, where the electric field force is large enough to balance the other forces.)

For nanometer-sized dust particles (see Fig. 3.6) gravity is negligible. Then, ion drag force becomes the dominant force which has to be balanced by the electric field force. The same is true for micron-sized dust under microgravity conditions where, of course, gravity does not play a role. Microgravity conditions can be realized in experiments on the International Space Station (ISS) [24, 25], on parabolic flights [26–32] or at drop towers.

Consequently, these relevant forces dominate the mechanism of particle trapping in plasma discharges. For large particles in the micrometer range where the dominant forces are gravity and electric field force a force balance is only obtained in the sheath of the lower electrode, where the upward electric field force is strong enough to balance the particles against gravity. Since the electric field in the sheath increases towards the electrode there typically is only a single position where electric field force and gravity balance. Thus, horizontally extended, but vertically restricted, flat dust arrangements are possible (see Figs. 3.7a and 3.8).

---

[6]For a 3 cm electrode gap this corresponds to a temperature difference of 6°C, only.

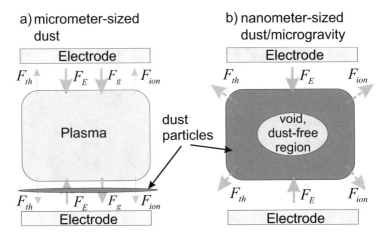

Fig. 3.7 Trapping and levitation of dust particles: (**a**) large particles, gravity is dominant, (**b**) small particles or microgravity conditions, gravity is unimportant

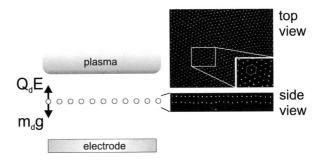

Fig. 3.8 Trapping and levitation of large dust particles where gravity is dominant. Here, an example of a dust cloud with two vertical layers is shown together with a top view of the particle arrangement

For smaller particles in the nanometer range or for large particles under microgravity conditions gravity is not important. Smaller electric fields are sufficient to levitate and trap the particles. Thus, particles can be trapped in the plasma bulk. In the plasma, the plasma potential is most positive in the center. Small ambipolar electric fields exist: the electric potential slowly drops to the sheaths that establish in front of walls and electrodes. Hence, in the plasma an outward pointing electric field exists that also drives the ambipolar, outward motion of the ions. As a consequence, the electric field force is pointing into the plasma bulk for the negatively charged dust particles, whereas ion drag (and also the thermophoretic force) point outward. Thus, trapping of dust particles should be possible in the entire plasma volume and three-dimensional extended dust clouds should be formed (see Fig. 3.7b). This is confirmed for submicron particles (grown in a silane discharge, see Sect. 11) and for large, micrometer-sized particles, under microgravity, see Fig. 3.9.

a)

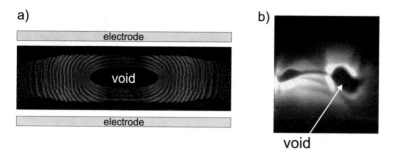

b)

void

Fig. 3.9 Trapping and levitation of (**a**) large dust particles (7.1 μm in diameter) under the weightlessness conditions of parabolic flights and (**b**) small, submicron, particles (grown in a silane discharge) for which gravity is unimportant. Here, individual particles cannot be resolved, the dust appears as bright cloud. In both cases a cross section of the dust cloud is illuminated by a laser and the scatter light is viewed with video cameras. In both situations, a central, dust-free void is formed

However, it is found that for these extended dust clouds large regions in the plasma center without dust particles, so-called "voids", exist, see Fig. 3.9. These voids are also due to the ion drag and electric field force [24, 33, 34]: in the plasma center, due to symmetry, the electric fields are small, and so is the ambipolar ion flow velocity. However, even at a slow ion flow, the ion drag force is already noticeable and larger than the electric field force (see Fig. 3.4).

Moreover, extended dust clouds (with voids) can be formed in the laboratory by balancing gravity by thermophoretic forces. These thermophoretic forces can be achieved by heating the lower electrode, see e.g. [17–19] and Sect. 8.4.

3.8 Vertical Oscillations

We have seen that large, micron-sized particles are trapped in the plasma sheath due to a balance of electric force and gravity. However, in the non-neutral, non-equilibrium plasma sheath, the charging processes are often difficult to describe when all processes like the oscillating rf-potentials, ion-neutral collisions and all the various charging currents are taken into account. Hence, to *measure* the charge of the dust particles vertical oscillations provide a useful and essential tool. In an additional step, nonlinear oscillations allow to extract information of the electric environment of the plasma sheath.

In the sheath where the particles are trapped the electric fields are generally spatially dependent, i.e. $E = E(z)$. The electric field usually increases monotonously from the plasma-sheath boundary, where the electric field is small, towards the electrode where it is largest. The dust charge will also be spatially dependent due to the space-dependent densities and velocities of electrons and ions in the sheath. Typically, a unique equilibrium position z_0 exists where the particles are trapped

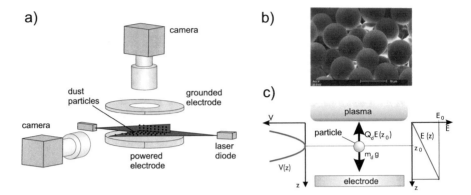

Fig. 3.10 (a) Scheme of the experimental setup in a typical experiment on dusty plasmas. The particles are illuminated by vertical and horizontal laser sheets. The particle motion is recorded from top and from the side with video cameras. (b) Electron micrograph of the MF particles typically used in the experiments. (c) Trapping of the particles in the sheath of an rf discharge. See text for details

(see Figs. 3.8 and 3.10c), i.e.

$$|Q_d(z_0)|\, E(z_0) = m_d g. \tag{3.19}$$

There, the upward electric force on the negative dust is balanced by gravity. Then, the equation of motion for a particle in the vertical direction (relative to the equilibrium position) is given by

$$\ddot{z} + \beta\dot{z} = g - \frac{|Q_d(z)|E(z)}{m_d} + \frac{F_{ext}}{m_d}, \tag{3.20}$$

where β is the friction coefficient describing the neutral gas drag [Eq. (3.14)] and F_{ext} are other external forces applied to the particle. To solve this equation of motion, one has to consider that the dust charge generally depends on the plasma conditions and therefore is itself a dynamic variable. In the following, we will investigate vertical oscillations of dust particles in the sheath for various situations.

3.8.1 Linear Resonance and Charge Measurement

In the most simple picture, vertical resonances can be treated in the following way: For small deviations from the equilibrium position the particle charge can be assumed to be spatially constant $Q_d(z) = Q_{d,0}$ and the electric field as linearly increasing $E(z) = E(z_0) + E_1(z - z_0)$. Here,

$$E_1 \equiv \frac{\partial E}{\partial z} = -\frac{\partial^2 \phi_{pl}}{\partial z^2} = \frac{e}{\varepsilon_0}(n_i - n_e)$$

is the slope of the electric field. Neglecting any horizontal variations, the slope of the electric field has been related via Poisson's equation to the electron and ion densities and the plasma potential ϕ_{pl}. When the charge density difference $n_i - n_e$ is constant (the so-called "matrix" sheath model) the slope of the electric field is constant. (see Fig. 3.10c). Such a linear electric field model is supported by a number of simulations of rf discharges [35] and theoretical analysis [36]. Under these assumptions the equation of motion reads (again z is the excursion from the equilibrium z_0)

$$\ddot{z} + \beta\dot{z} + \frac{|Q_{d,0}|E_1}{m_d}z = \frac{F_{ext}}{m_d}.$$

This equation is just that of a damped harmonic oscillator where the microspheres are trapped in the harmonic potential well [37, 38]

$$\frac{1}{2}m_d\omega_{res}^2 z^2 = \frac{1}{2}|Q_{d,0}|E_1 z^2 \tag{3.21}$$

with the resonance frequency of

$$\omega_{res}^2 = \frac{|Q_{d,0}|}{m_d}E_1. \tag{3.22}$$

So, the measurement of the vertical resonance of a trapped particle gives us a handle on the determination of the charge-to-mass ratio and allows to determine the particle charge $Q_{d,0}$, if the mass of the particles is known [37, 38].

The resonance frequency can be determined, e.g. from measuring the resonance curve by applying a periodic external force. The resulting resonance curve is known from the damped oscillator as

$$A(\omega) = A_0\frac{1}{\sqrt{(\omega_{res}^2 - \omega^2)^2 + \beta^2\omega^2}}, \tag{3.23}$$

where A_0/ω_{res}^2 is the oscillation amplitude of the particle for $\omega \to 0$.

The assumption of a constant space charge density $e(n_i - n_e) = \varepsilon_0 E_1$ models the linearly increasing electric field (with the slope E_1) in agreement with simulations of the rf sheath [35]. This allows the connection of the sheath electric field to the ion density measured by Langmuir probes in the bulk plasma. For a high voltage sheath of an rf sheath, the time averaged electron density $n_e = \alpha n_i$ is just the fraction α of the rf period where the particle "sees" a quasineutral environment due to the periodic rf oscillation of the plasma sheath, as discussed in Sect. 2.5.1. For the condition of micron-sized particles in an rf sheath a rough estimate is $\alpha \approx 1/3$ [38]. This results in the following form of the resonance frequency used to analyze the experiments

$$\omega_{res}^2 = \frac{|Q_{d,0}|n_i(1 - \alpha)e}{\varepsilon_0 m_d}. \tag{3.24}$$

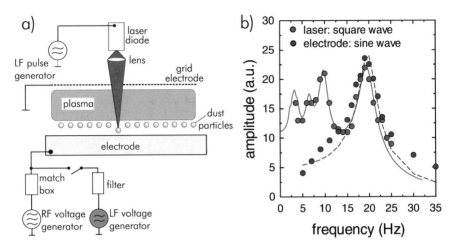

Fig. 3.11 Measuring the charge on MF microspheres. (**a**) Experimental setup for excitation of resonances by rf voltage modulation and laser manipulation. (**b**) Resonance curves obtained for a 9.47 μm MF particle for excitation by electrode voltage modulation and by laser pressure. From [40]

The charge measurements have been performed using monodisperse MF (melamine formaldehyde) microspheres (see Fig. 3.10b), which are perfectly spherical and have a well-defined mass. Therefore, the particles have the same charge and are trapped at the same height in the sheath. A number of vertical resonance measurements have been made using these particles [38, 39]. The vertical oscillations were driven by applying a very low-frequent modulation of the electrode rf voltage, see Fig. 3.11a. In doing so, the sheath width is modulated and the particle is forced to oscillate vertically in the trapping potential well.

From a frequency scan, a resonance in oscillation amplitude was obtained near 20 Hz for dust particles of $2a = 9.47$ μm diameter, see Fig. 3.11b. The measured data points are fitted with a resonance curve according to Eq. (3.23). From the fit the parameters A_0, the amplitude at very low frequencies, β and ω_{res} are obtained (Here: $A_0 = 5\omega_{res}^2$ a.u., $\beta = 26$ s^{-1} and $\omega_{res}/(2\pi) = 19.9$ Hz). From this using Eq. (3.22), the particle charge is found to be about 10,000 elementary charges and the floating potential is about 3 V. Estimations based on OML charging according to Eq. (2.19) result in charges that agree with the measured values within a factor of 2–3.

Applying this technique for plasmas at different pressures it is seen that the dust charge slowly decreases from about 11,000 to 6000 elementary charges with increasing pressure (40–120 Pa), see Fig. 3.12. This cannot be explained by pure OML charging, since the defining quantity, the electron temperature, is nearly constant ($T_e = 2.2$ eV) in that pressure range. Rather, it is the charge reduction due to the ion-neutral collisions as described in Sect. 2.4.4, see also [41].

Fig. 3.12 Measured dust charge as a function of discharge pressure. The uncertainty in the measured values is due to the uncertainty of the ion matrix sheath model, i.e. the uncertainty in the value α. From [39]

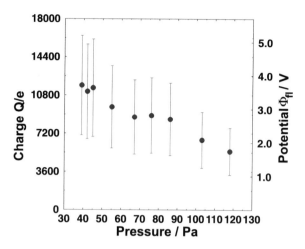

The width of the measured vertical resonance peak is determined by the neutral gas drag on the particle and is in quantitative agreement with the Epstein [13] friction coefficient β in Eq. (3.20). For the above mentioned experiment (at a gas pressure of 70 Pa) the expected friction coefficient is calculated to be in the range between $\beta = 20$ s^{-1} and 29 s^{-1}, depending on the parameter δ, the measured value is $\beta = 26$ s^{-1}.

An alternative, non-invasive technique to excite the dust particle resonance is by means of a focused laser beam as described in Sect. 3.6. Here, the laser beam is used to push the particle in the direction of the beam. By switching the laser "on" and "off" the vertical resonance curve of can be excited and measured (see Fig. 3.11) similar to the electrode manipulation. It is found that the resonance excited by the laser technique is nearly identical with the electrode manipulation. With laser excitation additional spurious resonances at $\omega_{res}/2$, $\omega_{res}/3$ etc. are excited due to the square wave excitation (laser "on" and "off") compared to the sinusoidal excitation at the electrode. Since the non-invasive laser technique and the electrode modulation give almost identical results it can be concluded that applying a small-amplitude potential modulation to the electrode does not lead to a severe disturbance of the plasma sheath environment.

Recently, the sensitivity of the vertical resonance method has been considerably improved by taking into account the phase relation between the exciting force F_{ext} and the particle oscillations [42]. This so-called phase-resolved resonance method allows to reveal subtle changes in the particle properties and particle interactions [15, 43].

3.8.2 Parametric Resonances

Besides the linear resonances, also parametric and nonlinear resonances have been observed in dusty plasmas. The nonlinear resonances will be described in the next section. Now, here, parametric resonances will be discussed. Parametric resonances occur when the confinement is periodically modulated. A mechanical paradigm for parametric resonances is the children's swing.

Parametric resonances in dusty plasmas can be excited by placing a wire in the plasma sheath close to the dust particles. When applying sinusoidal electrical potentials to the wire vertical oscillations of the dust particles are observed. The resulting resonance curve is shown in Fig. 3.13a for different excitation voltages. It is seen that for small voltages a single vertical resonance at ω_{res} as in the linear case is driven. At higher electrical amplitudes, however, a second resonance at $2\omega_{res}$ suddenly appears. This second resonance is a clear indication of parametric resonance. Thus, driving the particles at the double of the resonance frequency also leads to large vertical particle oscillations [44].

Parametric resonances occur when the strength of the external confinement is periodically modulated. The equation of motion then reads

$$\ddot{z} + \beta\dot{z} + \omega_{res}^2 \left(1 + h\cos\Omega t\right) = 0. \tag{3.25}$$

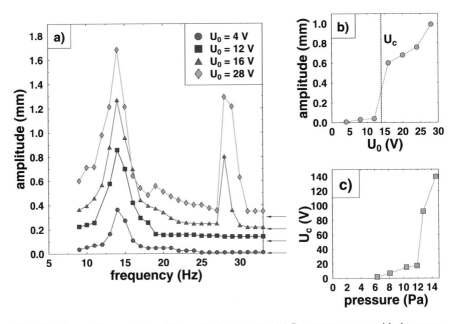

Fig. 3.13 Parametric excitation of the vertical resonance. (**a**) Resonance curves with the appearance of the second resonance at higher excitation voltages. (**b**) Amplitude of the second resonance versus excitation voltage and (**c**) critical excitation voltage as a function of gas pressure. From [44]

Here, the strength of the confinement changes periodically with a frequency Ω that generally is different from ω_{res} and h is the corresponding modulation depth. In our case, the resonance frequency ω_{res} is due to the confinement of the particle in the sheath by gravity and electric field force as discussed above. The modulation at Ω is due to the electrostatic potential on the wire. It can be suspected that the potential on the wire influences the electron and ion densities in the sheath that in turn modify the dust confinement. So, the periodic potential on the wire can periodically modulate the confinement. With increasing excitation voltage the modulation depth h increases.

Equation (3.25) is the famous Mathieu's equation in mechanics. It is known, that parametric resonances occur when the modulation frequency Ω is close to ω_{res} or $2\omega_{res}$. When friction is present (i.e. $\beta > 0$) the second resonance occurs only when the modulation depth exceeds a threshold value. This is also seen in Fig. 3.13b,c where a certain excitation amplitude is needed to excite the second resonance. This excitation threshold also increases with increasing discharge pressure, i.e. gas friction. Such a behavior is expected from a parametric oscillator.

Following that reasoning, the occurrence of parametric resonances means that the external confinement is modulated and disturbed by applying an electrostatic potential to the wire, a situation that is not observed by modulation of the rf voltage. Thus, wire excitation has to be used with great care in dusty plasma experiments.

3.8.3 Nonlinear Resonances

Finally, we like to turn to nonlinear vertical particle oscillations. As outlined in Eq. (3.20) a position-dependent charge and/or higher than linear terms in the electric field mean nonlinear modifications to the harmonic potential well. Obviously a non-harmonic potential well leads to nonlinear resonances. The electric field is only linear when the difference in charge densities of electrons and ions is constant (matrix sheath). If that does not hold the electric field will vary in a nonlinear way.

The dust charge in the sheath can be position-dependent due to two counteracting effects, see Fig. 3.14a. As an illustration, the charge is calculated using the OML model with streaming ions according to Eq. (2.12) together with a simple dc model of a space charge sheath to account for varying electron and ion densities and velocities. In the presheath the ions are accelerated to Bohm velocity. At the sheath edge ($z \approx 0.5$ mm) the ion velocity further increases by acceleration in the sheath electric field. This leads to a less efficient discharging of the dust grains by the ions and thus a higher (negative) potential, i.e. charge, on the particle. On the contrary, deep in the sheath the electron density is considerably reduced leading to smaller electron current and, hence, less negative dust particles, see also Fig. 2.3. Consequently, the particle first becomes more negative when moving into the sheath and again somewhat more positive deeper in the sheath.

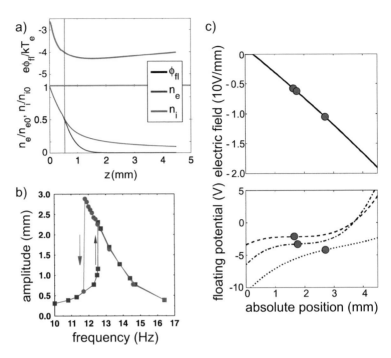

Fig. 3.14 (**a**) Calculated particle potential in the sheath using the OML model and a standard sheath model for the electron and ion densities and velocities. The vertical dashed line indicates where the presheath evolves into the actual sheath. (**b**) Measured nonlinear resonance curve for a 9.47 μm particle. Note the hysteresis in the resonance for increasing and decreasing frequency. (**c**) Linear electric field and position dependent particle potential profiles that will lead to the observed nonlinear resonance for three different particle sizes. From [45]

To analyze nonlinear oscillations, we assume that $Q(z)$ and $E(z)$ can be expanded in polynomials[7] according to

$$E(z) = E_0 + E_1 z + E_2 z^2 + E_3 z^3 \dots$$
$$Q(z) = Q_0 + Q_1 z + Q_2 z^2 + Q_3 z^3 \dots \quad (3.26)$$

Then, the equation of motion becomes nonlinear (up to 3rd order is considered, here, corresponding to a potential well of 4th order) with

$$\ddot{z} + \beta \dot{z} + C_1 z + C_2 z^2 + C_3 z^3 = \frac{F_{ext}}{m_d}. \quad (3.27)$$

---

[7]Here, for convenience, we have dropped the index "d" for the dust charge and use $Q(z) = |Q_d(z)|$ as positive.

The coefficients of nonlinearity are given by

$$C_1 = \frac{1}{m_d}(Q_0 E_1 + Q_1 E_0)$$

$$C_2 = \frac{1}{m_d}(Q_0 E_2 + Q_1 E_1 + Q_0 E_2)$$

$$C_3 = \frac{1}{m_d}(Q_0 E_3 + Q_1 E_2 + Q_2 E_1 + Q_3 E_1) \tag{3.28}$$

The potential well for the particles can then be written as

$$V(z) = \frac{1}{2}m_d C_1 z^2 + \frac{1}{3}m_d C_2 z^3 + \frac{1}{4}m_d C_3 z^4 \ldots \tag{3.29}$$

Here, $C_1 > 0$ corresponds to the (modified) linear resonance frequency, compare Eq. (3.21), $C_2 > 0$ or $C_2 < 0$ to an up/down asymmetry of the vertical potential well and $C_3 < 0$ or $C_3 > 0$ to a weakening/strengthening of the potential well with larger oscillation amplitudes. Nonlinear resonance curves have been excited in experiments at quite low gas pressure and high excitation amplitudes. In experiments of Ivlev et al. [46] the nonlinear oscillations were driven by applying a voltage to a wire placed below the dust particle. In contrast, Zafiu et al. [45] used the rf voltage modulation at the electrode.

The measured resonance curve, see Fig. 3.14b, shows distinct features of non-linearity. First, the resonance curve is not smooth. At certain excitation frequencies there is a jump in the oscillation amplitude from small to large values. Second, the resonance curve is bent towards lower frequencies and, third, there is a pronounced hysteresis in the resonance curve when going from smaller to larger frequencies or the opposite way.

The main resonance frequency is about 13 Hz, here, which corresponds to that of the linear resonance under these discharge conditions. The bending of the resonance towards lower frequencies is a clear indication that the potential well gets weaker with increasing amplitude, i.e. $C_3 < 0$. (For $C_3 > 0$ a bending towards higher frequencies would be expected.) This is easily seen from the following reasoning: Near the resonance the particle has a large oscillation amplitude and "feels" also the nonlinearity C_3. When the potential gets weaker with increasing amplitude, the particle feels weaker mean restoring forces and the resonance frequency is then shifted towards lower frequencies. At lowered frequencies then again higher oscillation amplitudes can be sustained that further decrease the resonance frequency. In this way, a bending towards lower frequencies is observed.

By comparing the measured resonance curve with calculated resonances using the equation of motion (3.27) the coefficients of nonlinearity can be determined. These coefficients can then be related to position dependent dust charges or electric field profiles that may range from linear via parabolic to cubic. When using different particle sizes different regions of the sheath can be probed resulting in a more or less consistent set of parameters Q_i and E_i. Zafiu et al. [45] found best agreement among experiments with different particle sizes for a linearly increasing electric field and a position dependent dust charge, see Fig. 3.14c. As mentioned above, the position dependent dust charge is due to the increased ion stream and the reduced electron density in the sheath. Position-dependent dust charges have been reported recently in experiments using hypergravity conditions [47].

3.8.4 Delayed Charging

The idea of a time and space dependent charge is supported by the observation of self-excited vertical oscillations, see Fig. 3.15. There vertical particle oscillations have been observed that grow in time without external drivers [48]. The growth time is of the order of 10 s. The oscillations reach a large amplitude until the particles drop from the discharge. Such oscillations are only possible if a source of energy is provided that can overcome the energy loss by friction with the neutral gas. A possible energy gain mechanism can lie in the combination of a position-dependent dust charge and finite charging times. When during the oscillation, due to finite charging times, the dust particle has an instantaneous charge $Q_t(z)$ which is different from the equilibrium charge $Q_{eq}(z)$ the restoring force $F = Q_t(z)E(z)$ and thus the restoring energy $\int F \, ds$ is different from the equilibrium situation. The difference is illustrated by the shaded area in Fig. 3.15. Careful analysis [46] of this delayed charging has shown that indeed such a mechanism could overcome the energy loss by gas friction at low gas pressures. These self-excited oscillations and the nonlinear resonances clearly identify the dust charge as a dynamical variable.

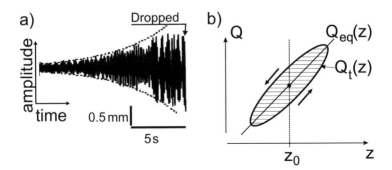

Fig. 3.15 (**a**) Self-excited vertical oscillations observed in a DC discharge. Reprinted figure with permission from [48]. © (1999) by the American Physical Society. (**b**) Illustration of the mechanism of delayed charging after [48, 49]

3.9   Other Charge Measurements

Other in-situ charge measurements make use of the particle-particle interaction by the analysis of wave propagation or particle-particle collisions. These methods allow or require a simultaneous measurement of the shielding length which also defines the interaction strength. These methods and their results will be presented in Sect. 4.1 and Chaps. 7 and 8, but it should already be mentioned here that the charges obtained from particle-particle interactions are very similar to those of the resonance method.

In ex-situ charge measurements [50] particles are dropped through a discharge into a Faraday cup where their charge is measured. Charged dust particles entering the Faraday cup lead to an electrical current I in the cup from which the dust charge is measured as $Q_d = \int I \, dt$. In their experiments the influence of different charging mechanisms like electron beams or UV radiation was investigated, see Fig. 3.16. Without any external influence the dust is slightly positively charged due to triboelectric charging when leaving the dust dropper. With UV radiation, the particles become more positively charged. When a photoemitting cathode is introduced, the dust is charged negatively due to the electrons released from the cathode.

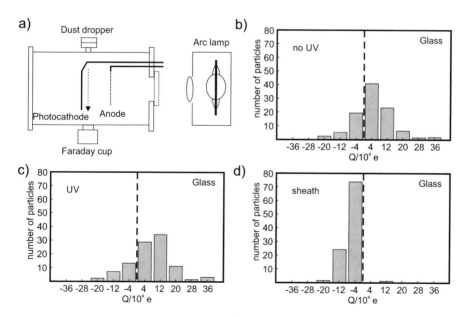

Fig. 3.16 (**a**) Scheme of the experimental setup. The particles fall from the dropper into the Faraday cup, where the dust charge is measured. The arc lamp provides UV radiation. (**b**) Dust charge distribution of 100 glass spheres without any UV irradiation, (**c**) with UV irradiation and (**d**) with a photoemitting cathode present. Reprinted figures with permission from [50]. © (2000) by the American Physical Society

References

1. S. Hamaguchi, R.T. Farouki, Phys. Rev. E **49**, 4430 (1994)
2. J.D. Jackson, *Classical Electrodynamics* (Wiley, New York, 1962)
3. I.H. Hutchinson, Plasma Phys. Control. Fusion **47**, 71 (2005)
4. S.A. Khrapak, A.V. Ivlev, V.V. Yaroshenko, G.E. Morfill, Phys. Rev. Lett. **102**, 245004 (2009). https://link.aps.org/doi/10.1103/PhysRevLett.102.245004
5. M.S. Barnes, J.H. Keller, J.C. Forster, J.A. O'Neill, D.K. Coultas, Phys. Rev. Lett. **68**, 313 (1992)
6. M.D. Kilgore, J.E. Daugherty, R.K. Porteous, D.B. Graves, J. Appl. Phys. **73**, 7195 (1993)
7. S. Khrapak, A.V. Ivlev, G. Morfill, H. Thomas, Phys. Rev. E **66**, 046414 (2002)
8. S.A. Khrapak, A.V. Ivlev, S.K. Zhdanov, G.E. Morfill, Phys. Plasmas **12**, 042308 (2005)
9. I.H. Hutchinson, Plasma Phys. Control. Fusion **48**, 185 (2006)
10. L. Patacchini, I.H. Hutchinson, Phys. Rev. Lett. **101**, 025001 (2008)
11. V.R. Ikkurthi, K. Matyash, A. Melzer, R. Schneider, Phys. Plasmas **16**, 043703 (2009)
12. I.H. Hutchinson, C.B. Haakonsen, Phys. Plasmas **20**, 083701 (2013). https://doi.org/10.1063/1.4818144
13. P.S. Epstein, Phys. Rev. **23**, 710 (1924)
14. B. Liu, J. Goree, V. Nosenko, L. Boufendi, Phys. Plasmas **10**, 9 (2003)
15. H. Jung, F. Greiner, O.H. Asnaz, J. Carstensen, A. Piel, Phys. Plasmas (1994-present) **22**, 053702 (2015). http://dx.doi.org/10.1063/1.4920968
16. H. Vestner, L. Waldmann, Physica **86A**, 303 (1977)
17. H. Rothermel, T. Hagl, G. Morfill, M. Thoma, H.M. Thomas, Phys. Rev. Lett. **89**, 175001 (2002)
18. C. Schmidt, O. Arp, A. Piel, Phys. Plasmas **18**, 013704 (2011)
19. O. Arp, D. Block, A. Piel, A. Melzer, Phys. Rev. Lett. **93**, 165004 (2004)
20. A. Ashkin, Phys. Rev. Lett. **24**, 156 (1970)
21. A. Melzer, Plasma Sources Sci. Technol. **10**, 303 (2001)
22. J. Schablinski, D. Block, A. Piel, A. Melzer, H. Thomsen, H. Kählert, M. Bonitz, Phys. Plasmas **19**, 013705 (2012)
23. H. Thomsen, H. Kählert, M. Bonitz, J. Schablinski, D. Block, A. Piel, A. Melzer, Phys. Plasmas **19**, 023701 (2012)
24. G.E. Morfill, H. Thomas, U. Konopka, H. Rothermel, M. Zuzic, A. Ivlev, J. Goree, Phys. Rev. Lett. **83**, 1598 (1999)
25. V.E. Fortov, O.S. Vaulina, O.F. Petrov, V.I. Molotkov, A.V. Chernyshev, A.M. Lipaev, G. Morfill, H. Thomas, H. Rothermel, S.A. Khrapak, Y.P. Semenov, A.I. Ivanov, S.K. Krikalev, Y.P. Gidzenko, JETP **96**, 704 (2003)
26. S. Khrapak, D. Samsonov, G. Morfill, H. Thomas, V. Yaroshenko, H. Rothermel, T. Hagl, V. Fortov, A. Nefedov, V. Molotkov, O. Petrov, A. Lipaev, A. Ivanov, Y. Baturin, Phys. Plasmas **10**, 1 (2003)
27. M.H. Thoma, H. Höfner, M. Kretschmer, S. Ratynskaia, G.E. Morfill, A. Usachev, A. Zobnin, O. Petrov, V. Fortov, Microgravity - Science and Technology **8**(3), 47 (2006). http://dx.doi.org/10.1007/BF02870378
28. M. Klindworth, A. Piel, A. Melzer, Phys. Rev. Lett. **93**, 195002 (2004)
29. M. Wolter, A. Melzer, O. Arp, M. Klindworth, A. Piel, Phys. Plasmas **14**, 123707 (2007)
30. M. Schwabe, S. Zhdanov, H. Thomas, A.V. Ivlev, M. Rubin-Zuzic, G.E. Morfill, V.I. Molotkov, A.M. Lipaev, V.E. Fortov, T. Reiter, New J. Phys. **10**, 033037 (2008)
31. B. Buttenschön, M. Himpel, A. Melzer, New J. Phys. **13**, 023042 (2011)
32. T. Bockwoldt, O. Arp, K.O. Menzel, A. Piel, Phys. Plasmas (1994-present) **21**, 103703 (2014)
33. J.L. Dorier, C. Hollenstein, A. Howling, J. Vac. Sci. Technol. A **13**, 918 (1995)
34. J. Goree, G. Morfill, V. Tsytovich, S.V. Vladimirov, Phys. Rev. E **59**, 7055 (1999)
35. P. Belenguer, J.P. Blondeau, L. Boufendi, M. Toogood, A. Plain, A. Bouchoule, C. Laure, J.P. Boeuf, Phys. Rev. A **46**, 7923 (1992)

36. E. Tomme, D. Law, B.M. Annaratone, J. Allen, Phys. Rev. Lett. **85**, 2518 (2000)
37. A. Melzer, T. Trottenberg, A. Piel, Phys. Lett. A **191**, 301 (1994)
38. T. Trottenberg, A. Melzer, A. Piel, Plasma Sources Sci. Technol. **4**, 450 (1995)
39. A. Melzer, A. Homann, A. Piel, Phys. Rev. E **53**, 2757 (1996)
40. A. Homann, A. Melzer, A. Piel, Phys. Rev. E **59**, 3835 (1999)
41. S.A. Khrapak, S.V. Ratynskaia, A.V. Zobnin, A.D. Usachev, V.V. Yaroshenko, M.H. Thoma, M. Kretschmer, H. Höfner, G.E. Morfill, O.F. Petrov, V.E. Fortov, Phys. Rev. E **72**, 016406 (2005)
42. J. Carstensen, H. Jung, F. Greiner, A. Piel, Phys. Plasmas **18**, 033701 (2011)
43. J. Carstensen, F. Greiner, A. Piel, Phys. Rev. Lett. **109**, 135001 (2012)
44. H. Schollmeyer, A. Melzer, A. Homann, A. Piel, Phys. Plasmas **6**, 2693 (1999)
45. C. Zafiu, A. Melzer, A. Piel, Phys. Rev. E **63**, 066403 (2001)
46. A. Ivlev, U. Konopka, G. Morfill, Phys. Rev. E **62**, 2739 (2000)
47. J. Beckers, T. Ockenga, M. Wolter, W.W. Stoffels, J. van Dijk, H. Kersten, G.M.W. Kroesen, Phys. Rev. Lett. **106**, 115002 (2011)
48. S. Nunomura, T. Misawa, N. Ohno, S. Takamura, Phys. Rev. Lett. **83**, 1970 (1999)
49. S. Takamura, N. Ohno, S. Nunomura, T. Misawa, K. Asano, in *Frontiers in Dusty Plasmas*, ed. by Y. Nakamura, T. Yokota, P. Shukla (Elsevier Science B.V., Amsterdam, 2000), pp. 337–344. https://doi.org/10.1016/B978-044450398-5/50045-2. http://www.sciencedirect.com/science/article/pii/B9780444503985500452
50. A.A. Sickafoose, J.E. Colwell, M. Horányi, S. Robertson, Phys. Rev. Lett. **84**, 6034 (2000). http://link.aps.org/doi/10.1103/PhysRevLett.84.6034

Chapter 4
Dust Particle Interaction

After having identified the charging process and relevant forces on the dust we
will now turn to the interaction among the dust particles. We will first discuss
the situation for particles in the same horizontal layer. The interaction between
particles at different vertical positions will be discussed afterwards. We will learn
that the interaction among dust particles trapped in the sheath of a discharge is
dramatically different in the horizontal and in the vertical direction due to the ion
streaming motion. The ions introduce a wakefield or ion-focus interaction that leads
to instabilities which also drive phase transitions in plasma crystals.

4.1 Interaction in the Horizontal Plane

A close inspection of a 2D plasma crystal trapped in the sheath of a plasma
discharge, see Figs. 1.3 or 3.8 reveals very interesting details. In the horizontal
plane, the particles arrange in a hexagonal structure which is a minimum energy
configuration in monolayer for mutually repelling particles, see Chap. 5. A natural
choice for the interaction between particles in a plasma would be a (shielded)
Coulomb or Debye-Hückel potential. This interaction potential, also named Yukawa
potential in the field of complex fluids, is given as

$$\phi(r) = \frac{Q}{4\pi \varepsilon_0 r} \exp\left(-\frac{r}{\lambda_D}\right), \tag{4.1}$$

where λ_D is the linearized Debye shielding length

$$\lambda_D = \left(\frac{1}{\lambda_{D,e}^2} + \frac{1}{\lambda_{D,i}^2}\right)^{-1/2} \tag{4.2}$$

© Springer Nature Switzerland AG 2019
A. Melzer, *Physics of Dusty Plasmas*, Lecture Notes in Physics 962,
https://doi.org/10.1007/978-3-030-20260-6_4

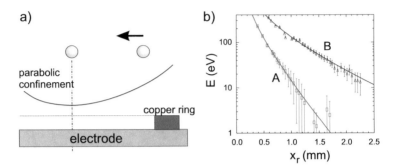

Fig. 4.1 Konopka's two-particle collision experiment: (**a**) Scheme of the experimental arrangement, (**b**) electrostatic interaction energy as a function of relative particle distance x_r derived from the experiment. Symbols denote experimental results, the lines the best fits of a Debye-Hückel (Yukawa) potential. The two curves A and B correspond to different discharge conditions (see text). After [1], data kindly provided by U. Konopka

with

$$\lambda_{\mathrm{De,i}} = \sqrt{\frac{\varepsilon_0 k_{\mathrm{B}} T_{\mathrm{e,i}}}{n_{\mathrm{e,i}} e^2}} \tag{4.3}$$

being the electron and ion Debye length, respectively. The Debye-Hückel interaction energy between two like-charged dust particles then is

$$V(r) = Q\phi(r) = \frac{Q^2}{4\pi \varepsilon_0 r} \exp\left(-\frac{r}{\lambda_{\mathrm{D}}}\right). \tag{4.4}$$

A direct experimental determination of the horizontal interaction potential V between microspheres has been achieved by Konopka et al. [1] who have studied central collisions between two particles. In this experiment, two microspheres ($2a = 8.9\,\mu\mathrm{m}$) were trapped in the sheath of an rf discharge. A copper ring was placed onto the electrode, compare Chap. 8. The copper ring leads to an additional horizontal parabolic confinement (of strength ω_0) which was confirmed by the analysis of the motion of a single particle. Now, one of the two microspheres was positioned at rest in the center of the confining potential, the other was pulled outwards ("up" the horizontal confinement) by a positively biased probe wire. This particle was subsequently released and collided with the central dust particle (see Fig. 4.1a). The collision dynamics has been analyzed to retrieve the particle interaction.

The equation of motion for the two particles reads (using the relative coordinate x_r and the reduced mass $\mu = m_1 m_2/(m_1 + m_2) = m_{\mathrm{d}}/2$)

$$\mu \ddot{x}_r + \mu \beta \dot{x}_r + \mu \omega_0^2 x_r = -\frac{\partial V(x_r)}{\partial x}, \tag{4.5}$$

where V is the interaction potential to be determined. The integration of the equation of motion gives

$$V(x_r) = \frac{1}{2}\mu\dot{x}_r^2 + \frac{1}{2}\mu\omega_0^2 x_r^2 + \mu\beta \int_0^t \dot{x}_r^2(\tau)d\tau, \quad (4.6)$$

where the first term is the kinetic energy of the two-particle system (using the reduced mass), the second describes the potential energy in the (weak) horizontal parabolic confinement and the third accounts for losses by friction with the neutral gas. The interparticle potential is measured from the relative particle positions x_r and velocities $\dot{x}_r$ which are easily determined from video data. This enables to reconstruct the interaction potential with good accuracy from the particle trajectories.

The experiments [1] show that the so obtained electrostatic energy directly reflects the interaction potential between the microspheres can be described very accurately by a Debye-Hückel (Yukawa) type interaction

$$V(x_r) = \frac{Z_d^2 e^2}{4\pi\varepsilon_0 x_r} \exp\left(-\frac{x_r}{\lambda_D}\right),$$

as shown in Fig. 4.1b, where a Debye-Hückel interaction energy was fitted to the experimental results. From the fit the following values of the charge and shielding length have been obtained [1]: $Z_d = 13\,900$, $\lambda_D = 0.34$ mm (for case A at high discharge voltage $U_{pp} = 233$ V), $Z_d = 17\,100$, $\lambda_D = 0.78$ mm (case B, low voltage $U_{pp} = 64$ V). As expected the Debye length increases for lower plasma density (i.e., lower discharge voltage). The particle charges are in the range of the expected values from the charging theory and are comparable to the measurements using the resonance technique. The screening length is of the order of the electron Debye length in these discharges. This might be expected. However, that point has to be discussed in some more detail later in Sect. 7.9.

The horizontal interaction can also be derived from wave experiments. The dispersion relation of waves contains the particle-particle interaction. Wave dispersion, and thus particle interaction, will be discussed later in Chaps. 6 and 7.

4.2 Vertical Interaction

When returning to the plasma crystal in Figs. 1.3 or 3.8, the side view reveals that the particles of different vertical layers are located directly atop of each other. They are vertically aligned. This unusual ordering is practically always observed for two or more vertical layers and has been recognized early after the discovery of the plasma crystal [2–4]. The vertically aligned strings move as an entire chain in the crystal. This demonstrates that in the vertical direction additional attractive forces must be present. Possible sources and mechanisms for the formation of additional attractive forces will be described now.

Clearly, the observed vertical alignment of the microspheres certainly is not a minimum energy configuration for purely repulsive particle interactions. The fundamental reason for the difference between horizontal and vertical interactions lies in the fact that the main external forces on the microspheres—electric field force and gravity—act in the vertical direction. Due to gravity the microspheres are trapped in the non-neutral, non-equilibrium environment of the sheath which strongly influences the interaction between the microspheres.

For the very large and rod-like particles used in the experiment of Mohideen et al. [5] and Annaratone et al. [6] dipole moments on the particles might explain the observation of particle alignment along the electric field. However, these dipole effects are much too small to overcome the repulsion between the particles used in the experiments, here (see also Sect. 3.2.2). Hence, other mechanisms have to be taken into account.

The streaming ions in the sheath are the most obvious candidate as the source for the generation of attractive forces. Different mechanisms of ion-mediated attractive forces have been put forward in a number of theoretical models and simulations. Here, we will now discuss the formation of attractive forces from the viewpoints of a "wave" model and a "particle" model which reflect two sides of the same coin.

4.2.1 The Wakefield Potential (Wave Model)

In this class of models, the attraction is described by wakefield formation downstream a dust particles in a (supersonic) ion flow [7]. Here, it is assumed that the dust particle excites ion-acoustic waves of all frequencies in the ion stream. The ion-acoustic waves superpose to form an (oscillating) potential structure. The regions with positive potential provide an attractive force for another (negatively charged) dust particle.

In this approach the interaction between the ion flow and a single dust particle is described with the help of the dielectric function ε. Here, the ion flow velocity u_i is assumed to be in the (vertical) z direction with $M = u_i/v_B$ being the Mach number of the flow ($v_B = \sqrt{k_B T_e/m_i}$ is the Bohm velocity). Since the dust particles are mainly trapped in the sheath a supersonic ion flow, $M > 1$, can be assumed.

The electrostatic potential around the dust particle is given from linear response theory in general terms by Lampe et al. [8]

$$\phi(\mathbf{r}) = \int \frac{Q_d}{8\pi^3 \varepsilon_0 q^2} \frac{1}{\varepsilon(\mathbf{q}, \omega - q_z u_i)} e^{i\mathbf{q}\cdot\mathbf{r}} d\mathbf{q}, \qquad (4.7)$$

where $\mathbf{q}$ ($q = |\mathbf{q}|$) is the wave vector of the excited ion acoustic waves and $\varepsilon(\mathbf{q}, \omega)$ is the dielectric response of the plasma. Equation (4.7) is just the Fourier notation of the particle potential and the dielectric function describes the response of the plasma species (electrons and ions) to the electrostatic potential. Here, we use a moving frame with the velocity of the streaming ions. The frequencies are therefore Doppler shifted $\omega \to \omega - q_z u_i$ (q_z is the z component of the wave vector).

The plasma dielectric response is given by

$$\varepsilon(\mathbf{q}, \omega - q_z u_i) = 1 + \frac{1}{q^2 \lambda_{D,e}^2} - \frac{\omega_{p,i}^2}{(\omega - q_z u_i)^2}, \tag{4.8}$$

where the second term on the RHS describes the electron shielding and the third term is the ion response. For the electrons the low-frequency limit of the dielectric response is used due to the high mobility of the electrons. For the ions the (Doppler shifted) high frequency limit is taken.

After some algebra the inverse of the dielectric function is found as

$$\frac{1}{\varepsilon(\mathbf{q}, \omega - q_z u_i)} = \frac{q^2 \lambda_{D,e}^2}{1 + q^2 \lambda_{D,e}^2} \left[1 + \frac{\omega_s^2}{(\omega - q_z u_i)^2 - \omega_s^2} \right], \tag{4.9}$$

where $\omega_s = q v_B / (1 + q^2 \lambda_{D,e}^2)^{1/2}$ is the frequency of oscillations in the ion flow. For comparison, the dielectric response of a Coulomb potential with screening by electrons (i.e. a Debye-Hückel potential) is simply given by

$$\frac{1}{\varepsilon(\mathbf{q}, \omega)} = \frac{q^2 \lambda_{D,e}^2}{1 + q^2 \lambda_{D,e}^2}.$$

By substituting the dielectric function (4.9) in Eq. (4.7) the resulting total potential can be written as the sum of two potentials

$$\phi(\mathbf{r}) = \phi_D(\mathbf{r}) + \phi_c(\mathbf{r}) \tag{4.10}$$

where

$$\phi_D(\mathbf{r}) = \frac{Q_d}{4\pi \varepsilon_0 r} \exp\left(-\frac{r}{\lambda_{D,e}}\right)$$

is the usual Debye shielding potential and

$$\phi_c(\mathbf{r}) = \int \frac{Q_d}{8\pi^3 \varepsilon_0 q^2} \frac{q^2 \lambda_{D,e}^2}{1 + q^2 \lambda_{D,e}^2} \frac{\omega_s^2}{(\omega - q_z u_i)^2 - \omega_s^2} e^{i\mathbf{q}\cdot\mathbf{r}} d\mathbf{q}. \tag{4.11}$$

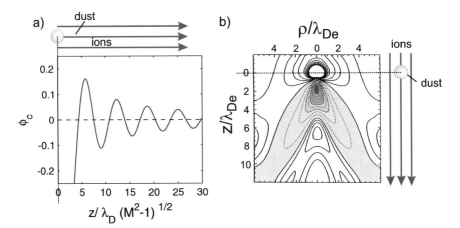

Fig. 4.2 (**a**) The ion wake potential in an ion flow. Downstream of the particle an oscillating potential is formed with alternating attractive ($\phi_c > 0$) and repulsive ($\phi_c < 0$) regions. After [7]. (**b**) Spatially resolved wakefield in the direction of the ion flow (z) and perpendicular to the flow (ρ). The potential contours are indicated (solid lines: negative potentials, dashed lines: positive potentials). Adapted and reprinted from [8], with the permission of AIP Publishing. The shaded area has a positive potential and is thus attractive for a second negative dust particle. Note the different axis scaling of z in panels (**a**) and (**b**). Note also the different orientation of the flow in (**a**) and (**b**)

Using cylindrical coordinates $(x, y, z) \rightarrow (\rho, \varphi, z)$, an approximate solution on the vertical axis $\rho = 0$, i.e. directly downstream the dust particle, is given by Vladimirov and Nambu [7]

$$\phi_c(\rho = 0, z) = \frac{Q_d}{4\pi\varepsilon_0 z} \frac{2\cos\left(z\,\lambda_{D,e}\sqrt{M^2 - 1}\right)}{1 - M^{-2}}. \qquad (4.12)$$

According to this model an oscillating ion wake potential ϕ_c downstream of the dust particle is created with an alternating sequence of regions with enhanced positive and negative potential (see Fig. 4.2a). For $z < 5\lambda_{D,e}\sqrt{M^2 - 1}$ the potential is negative due to the presence of the negative dust particle at $z = 0$. The potential (and the corresponding ion density) then "overshoots" and forms an attractive potential between $5 < z/\lambda_{D,e}\sqrt{M^2 - 1} < 8$. This wake potential is attractive to a second negatively charged particle since there $\phi_c > 0$. This ion wake may provide the attractive force necessary to explain the vertical ordering of the particles. This mechanism would be similar to Cooper pairing in superconductors [9], in that the dust particle polarizes the surrounding medium, the plasma, which in turn leads to attraction of other particles. In addition, it is reasoned that other dust particles will arrange in the areas of positive potential defined by the first particle. The vertical scales introduced by this collisionless model are different from those in the experiment: The model predicts that the lower particle would be found in the maximum of ϕ_c which is approximately at $z = 6\lambda_{D,e}\sqrt{M^2 - 1} \approx 3000$ µm when $\lambda_{D,e} = 500$ µm is assumed. This is far from the observed vertical distances of $d = 400$ to 600 µm.

Using a more advanced dielectric function including collisions (with the ion-neutral collision frequency ν_{in}) and Landau damping (through the application of a shifted Maxwellian velocity distribution f_{i0}) via

$$\varepsilon(\mathbf{q}, \omega - q_z u_i) = 1 + \frac{1}{q^2 \lambda_{D,e}^2} - \frac{\omega_{p,i}^2}{q^2} \int \frac{\mathbf{q} \, \partial f_{i0}(\mathbf{v})/\partial \mathbf{v}}{q_z u_i - \omega - i\nu_{in}} d\mathbf{v}, \tag{4.13}$$

a more realistic wakefield potential is derived [8]. This wakefield potential is shown in Fig. 4.2b spatially resolved along and transverse to the ion flow. Compared to Fig. 4.2a, the attractive potential maximum is shifted towards the particle and is now found at about $z \approx 2\lambda_{D,e}$ which agrees more closely with the experiment. The Landau and collisional damping also leads to a rapid decay of the oscillations along the flow so that generally only a single potential maximum behind the dust particle is obtained. Further, the wake is V-shaped similar to a Mach cone of the ion acoustic waves excited by the grain.

The above discussed wave model included the excitation of linear waves by a single dust particle. In a region near the dust the linear description might become invalid and effects introduced by the presence of a second particle in the wakefield are not addressed, here. It should be noted that wakefields are also found to form in subsonic ion flows where $M < 1$, see e.g. [10, 11].

4.2.2 The Ion Focus (Particle Model)

The problem of attraction will now be investigated on the microscopic particle (ion) level [12, 13]. For this purpose, the experimental conditions are closely mimicked in simulations: a two-layer plasma crystal in the sheath of an rf discharge is considered; the electric field and electron distribution are taken into account as time-averaged, but spatially dependent quantities.

Then, ion trajectories are calculated according to the following equation of motion in the electric field of the sheath $E_{sh}(z)$ and of the dust particles

$$m_i \ddot{\mathbf{r}} = e E_{sh}(z) - e \sum_j \nabla \phi_{d,j}(\mathbf{r}),$$

where $\phi_{d,j}(\mathbf{r})$ is the electric potential of the dust particle j on the ions according to Eq. (4.1) and the sum is over all dust particles in the upper and lower layer. In the ion motion also ion-neutral collisions are taken into account. This is done the following way: The above equation of motion is solved between the collisions of the ions. The time between two collisions is chosen randomly in such a way that a

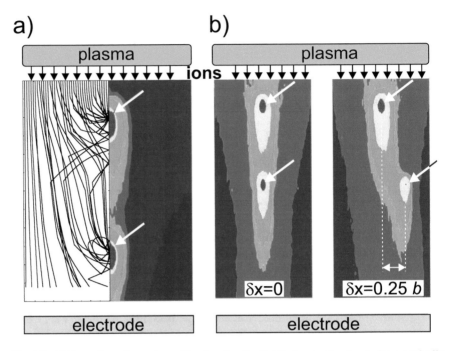

Fig. 4.3 (**a**) Ion trajectories and resulting ion density distribution in the sheath for a vertically aligned pair of microspheres. Charge exchange collisions are included in this simulation as can be seen by abrupt changes in the particle trajectories. (**b**) Ion density distribution for vertically aligned pairs (left, same as (**a**), but without the trajectories) and horizontally shifted lower particle (right). The particle positions are indicated by the arrows. After [12, 14]

defined ion mean free path ℓ_{mfp} is ensured (see Fig. 4.3a). In the collision the ion loses its *kinetic* energy and starts again with a random *thermal* velocity. This is done until the ion hits the electrode or a dust particle. It is also assumed that the ions enter the plasma sheath with Bohm velocity.

The advantage of following the ion trajectories is that from the principle of *actio = reactio* the force exerted on the ion by the dust particle is the same force that acts on the dust by the ions. Thus the force on the dust due to the ions can be directly determined from the ion trajectories. Similar refined and improved calculations have been performed by, e.g. [15–19].

This experiment-related approach ensures that the forces on the particles can be calculated under very realistic conditions. Figure 4.3a shows (a small fraction of) the simulated ion trajectories. From averaging over many trajectories, the ion density distribution is derived. The ion density distribution has been determined for different situations, see Fig. 4.3b: in the first case shown, the particles of the lower layer are exactly vertically aligned with the upper particles at a vertical distance (360 μm) taken from the experiment. Thus, here, the horizontal displacement of the lower layer is $\delta x = 0$. In the second case shown, the lower layer is displaced by a quarter of the horizontal interparticle distance ($\delta x = 0.25b$ with $b \approx 450$ μm).

It is seen that the shielding ion cloud around each particle is somewhat distorted from a spherical cloud and is extended downstream due to the ion flow. Looking at the individual trajectories, the ions are deflected into a region below the dust by the Coulomb collisions with the dust particles. That results in a region of enhanced ion space charge density ("ion focus") below each dust grain. This positive space charge is the reason for the attractive force on the particles. In this approach only a single attraction region is found in agreement with the more sophisticated wave model [8].

Both, the wakefield (wave model) and the ion focus (particle model) describe the formation of attractive regions downstream of a dust particle due to the interaction with streaming ions. The two models can be considered as complementary descriptions of the same phenomenon. However, ion trajectory simulations allow to reveal the particle interaction in more detail since the corresponding forces on the dust can be retrieved self-consistently.

4.3 Non-Reciprocal Attraction

Thus, such ion trajectory simulations reveal an extremely important and surprising phenomenon: the non-reciprocity of the attractive forces among two particles trapped at different heights in the sheath [12].

When looking at the ion density distributions in Fig. 4.3b one sees that, surprisingly, the ion cloud of the upper particles is not affected by the position of the lower particle. The upper half of Fig. 4.3b looks the same independent of whether the lower particle is directly beneath the upper particle or whether it is shifted.

In a somewhat simplified picture the behavior can be described as follows: Due to the supersonic ion flow the information on the location of the lower particles cannot be conveyed upstream to the upper particle. Hence, the ion motion and the ion focus are unaffected by the lower particles. So, since there is no reaction of the upper particles on the position of the lower, the interaction between the particles is non-reciprocal: *the lower particles are attracted by the ion focus of the upper, but there is no attractive counter-force acting on the upper particles.* This is an apparent violation of Newton's third law *actio = reactio* when looking only at the dust particles. Although each single ion reacts with the dust particles by *actio = reactio* the collective ion population does not. This is an outcome of the supersonic ion flow that introduces a broken symmetry into the system. In other words, the steady ion flow puts energy into the system, thus, it is an *open* system where Newton's third law does not necessarily hold [12, 20].

It should be made clear that this nonreciprocal attraction pertains essentially only to the *horizontal* force component for the lower and upper particle. The *vertical* forces between the two dust particles are found always repulsive, see also [21]. The interaction between particles at different heights must therefore be described by a horizontal non-reciprocal attractive force.[1]

---

[1]For a detailed quantitative analysis, one also could account for the different charging of the lower-layer particle inside and outside the ion focus [15–18, 21].

Due to the non-reciprocity of the interaction, we cannot assign a potential to this kind of interaction. The forces are non-conservative. Moreover, all techniques that rely on energy arguments cannot be directly applied to this situation. However, for a statistical mechanics approach under these conditions see [20].

To be more quantitative, the strength of the horizontal forces acting on the upper and lower particle has been calculated directly from the ion trajectories [12], see also [15–18]. Since we are mainly interested in the horizontal force component F_x that moves a lower-layer particle into the aligned position this horizontal force has been calculated from the ion trajectories as a function of the horizontal displacement δx of the lower-layer particles and is shown in Fig. 4.4a. It is seen that the horizontal attractive force on the lower particle from the ion focus is decisively larger than the (horizontal force component of the) repulsion from the upper particle. This means that, for the lower particle, the attraction overcomes the repulsion and drives the lower layer particle into the vertically aligned position which is then favored by the acting forces (remember: the upper particle does not feel any attraction, but only repulsion from the lower). It is also seen that the attractive force is linear for small displacements δx, i.e.

$$F_x = -k_{\mathrm{att}}\,\delta x,$$

which allows a linear stability analysis of the entire system (which is done in Sect. 4.5).

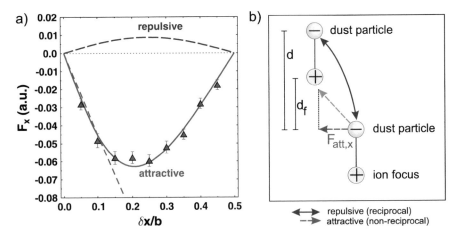

Fig. 4.4 (**a**) Strength of the attractive force (horizontal component) as a function of horizontal displacement. Symbols are results from the simulation, the solid line is from the model where the ion density distribution is replaced by a positive point charge. The dashed line indicates the linear behavior for small displacements. The repulsive force from the upper layer is shown for comparison. From [14]. (**b**) "Schweigert" model of the particle interaction derived from the simulation in Fig. 4.3. The ion focus is mimicked by a positive point charge beneath a dust particle. The attractive force acts only on the lower particle. From [12]

This interaction can be translated into the "Schweigert" model presented in Fig. 4.4b. The distorted ion cloud around the dust, i.e. the ion focus, can be mimicked very accurately by replacing the ion cloud with a positive point charge Z_f at a distance $d - d_f$ directly below the upper particle. Typical values are $Z_f = 0.5Z_d$ and $d_f = 0.4b$. The force exerted by this point-charge ion-focus exactly matches the forces determined from the full simulation, see Fig. 4.4a. The lower particles then feels the horizontal repulsion from the upper layer particle *and* the larger horizontal attraction from the ion-focus point charge below the upper particle. In turn, the upper particle only feels the repulsion from the lower and there is no attraction on the ion-focus point charge from the lower particle (remember: the ion cloud around the upper dust particle is not influenced by the lower particle).

4.4 Measurement of the Attractive Force

This peculiar type of particle interaction can be verified, also quantitatively, from experiments [22–26]. For example, the non-reciprocal attractive forces can be probed experimentally in a very simple experiment [23]. For this purpose, only two single particles are immersed into the plasma (see Fig. 4.5a). The first, upper has a smaller diameter and mass than the second, lower, particle. Due to their different masses the particles are trapped at different heights in the sheath, where the force balance is fulfilled for each of the particles. Thus, the vertical position of the particles is practically fixed. In this particular experiment, the upper particle had a diameter of 3.47 μm (and a mass of $m_1 = 3.31 \times 10^{-14}$ kg) and the lower one of 4.81 μm ($m_2 = 8.82 \times 10^{-14}$ kg).

In the horizontal plane, the microspheres can move freely under the influence of their mutual interaction. From the response of one particle to the motion of the other particle the interaction between them can be directly extracted.

As in the case of the excitation of the vertical resonance, the focused beam of a manipulation laser (690 nm, 40 mW) is used to push the upper and lower particle individually and in a defined manner.

Without any laser force applied the two single microspheres are found vertically aligned as in the case of a plasma crystal. When now the upper particle is pushed by the radiation pressure of the laser beam *both* upper *and* lower particle move together, their horizontal position is (nearly) identical, i.e. the particles stay vertically aligned (see Fig. 4.5b). That proves that the lower particle experiences a horizontal attractive force mediated by the upper particle.

When, however, the lower particle is pushed the two particles are separated horizontally, the alignment is broken (see Fig. 4.5c). The lower particle can be pushed far away from the upper ($t = 1 - 4$ s). When the laser is switched off again, the lower particle approaches the aligned position below the upper one, whereas the upper (although being the lighter particle) does not move towards the lower. In contrast, shortly before the lower particle reaches the aligned position (around $t = 6$ s) the upper particle experiences the repulsive force from the lower and moves

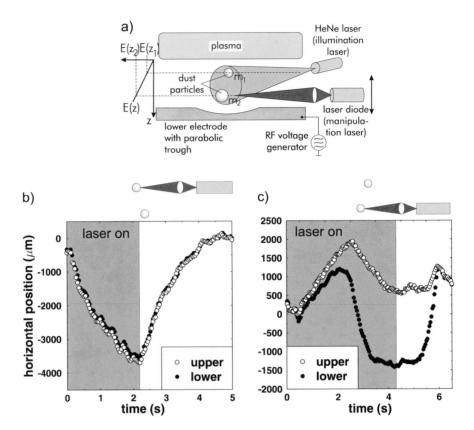

Fig. 4.5 Experiment on the non-reciprocity of the attractive force between two dust particles. (**a**) Scheme of the experimental setup. Horizontal position of upper and lower particle when (**b**) the upper particle is pushed and (**c**) when the lower particle is pushed. From [23, 24]

away from the lower. This definitely demonstrates that the upper particle does not feel an attractive force mediated by the lower.

This confirms the fact that for the horizontal interaction between particles at different heights Newton's third law *actio = reactio* seems to be violated under these conditions. This is an outcome of the supersonic ion flow that introduces a preferred direction into the system. In addition, the ion stream introduces a source of free energy, thus the dust particle system and its environment is an open system where simple Hamiltonian dynamics does not hold (again, see also [20]). One of the main outcomes in connection with the theoretical analysis presented is that the interaction force between the dust particles is non-reciprocal.

From the laser interaction with the two-particle dust system the horizontal attractive force component can also be derived quantitatively, see Fig. 4.6. By pushing the lower particle with the laser, the horizontal attractive force is balanced by the radiation pressure of the laser beam. By calibrating the laser force F_L using experiments on single particles we have determined the attractive force

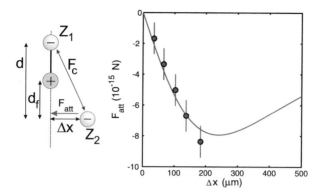

Fig. 4.6 Measurement of the attractive force between two dust particles. The measured magnitude of the attractive force (symbols) allows to derive the strength of the ion focus. The solid line is the calculated attractive force according to Eq. (4.14). After [24]

quantitatively. With increasing laser force the lower particle is shifted horizontally by a displacement Δx away from the aligned position.

Using the model of the positive point charge that mimics the ion focus the attractive force on the lower particle can be calculated. The horizontal component of the Coulomb force between the lower dust particle with charge Z_2 and the ion focus with charge Z_f is given by

$$F_{att} = -\frac{Z_f Z_2 e^2}{4\pi \varepsilon_0 r_{f2}^2} \frac{\Delta x}{r_{f2}} = -\frac{Z_f Z_2 e^2 \Delta x}{4\pi \varepsilon_0 (\Delta x^2 + d_f^2)^{3/2}}, \quad (4.14)$$

where $r_{f2} = (\Delta x^2 + d_f^2)^{1/2}$ is the distance between the ion focus and the lower particle. In Fig. 4.6 the measured forces are shown in comparison to the above equation. The charge on the upper and lower particle was measured as $Z_1 = 2200$ and as $Z_2 = 5900$, respectively, from the resonance method. Adjusting the values of Z_f and d_f in Eq. (4.14) agreement is found for $Z_f = 0.8 Z_1$ and $d_f \approx 400$ μm. These values are in good agreement with the simulations.

For small displacements Δx the force is linear with

$$F_{att} \approx -\frac{Z_f Z_2 e^2}{4\pi \varepsilon_0 d_f^3} \Delta x = -k_{att} \Delta x, \quad (4.15)$$

assuming $\Delta x \ll d_f$ in the denominator. Such effective "spring" constants k will be used in the stability analysis in the following section.

This "Schweigert" model [12] of the ion focus with the non-reciprocal attraction due to the formation of the ion focus has hence been verified qualitatively and quantitatively by experiments. Moreover, this model is also able to explain the stability of plasma crystals and phase transitions from the ordered, solid phase to an unordered, fluid phase [12, 14]. Moreover, this model has also been successfully applied to study mode-coupling instabilities [27].

4.5   Oscillatory Instability of the Vertical Alignment

After having shown that the horizontal component of the interaction between the microspheres at different heights is indeed given by a non-reciprocal attractive force the question arises under which conditions the vertically aligned plasma crystal observed in the experiment is stable or unstable. Here, this analysis is performed for the simplest case of a two-layer crystal with vertically aligned particles, see Fig. 4.7. There, the (reciprocal) repulsive forces between the negative dust particles and the (non-reciprocal) attractive forces on the lower layer particles towards the ion focus of the upper particles are indicated. From such a plasma crystal model the equations of motion in the horizontal plane for the nth particle in the upper (index 1) and lower (index 2) layer are

$$\ddot{x}_1^{(n)} + \beta \dot{x}_1^{(n)} = \frac{k_b}{m_d}\left(x_1^{(n-1)} - 2x_1^{(n)} + x_1^{(n+1)}\right) + \frac{k_d}{m_d}\left(x_1^{(n)} - x_2^{(n)}\right) \qquad (4.16)$$

$$\ddot{x}_2^{(n)} + \beta \dot{x}_2^{(n)} = \frac{k_b}{m_d}\left(x_2^{(n-1)} - 2x_2^{(n)} + x_2^{(n+1)}\right) + \frac{k_d}{m_d}\left(x_2^{(n)} - x_1^{(n)}\right)$$
$$- \frac{k_{f0}}{m_d}\left(x_2^{(n)} - x_1^{(n)}\right) - \frac{k_{f1}}{m_d}\left(x_1^{(n-1)} + x_1^{(n+1)} - 2x_2^{(n)}\right). \qquad (4.17)$$

Here, a small horizontal elongation x from the vertically aligned position is assumed. In addition, d and d_f are the vertical distance between the lower and upper layer and lower layer and ion focus, respectively, and Z_f the (positive) charge of the ion focus.

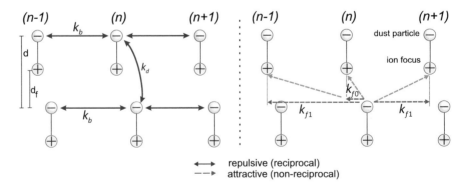

Fig. 4.7 Forces in a two-layer plasma crystal. Left: Reciprocal repulsive forces between all particles. Right: Additional horizontal non-reciprocal attraction on the lower layer particles

The first term on the RHS is the (linearized) repulsive interaction between the particles of the same layer (upper or lower). We characterize the interaction by the spring constant k_b for particles at a horizontal equilibrium distance b (see also Sect. 7.1). This determines the horizontal oscillation frequency ω_b of the particles in the respective plane (see below). The second term is the repulsion between upper and lower aligned particle (both taken as pure Coulomb forces for simplicity, here, denoted by k_d). The attraction by the ion focus at the same horizontal position (n) appears only in the equation of motion for the second layer due to the non-reciprocity of the attractive force as effective spring k_{f0}. Finally, the interaction of the second-layer dust particle (n) with the ion foci at $(n-1)$ and $(n+1)$ are taken into account as k_{f1}. The model presented here is simplified, it has, however, all the necessary ingredients to describe the physical mechanisms and to explain the experimental findings. A complete analysis taking into account a full 2D two-layer plasma crystal with hexagonal order and screened interaction is somewhat more difficult [12], but also does not include significant differences.

4.5.1   Solution of the Coupled Oscillations

To solve the equation of motion, first, we take the usual solution for waves on a linear chain

$$x_j^{(n)} = x_k \exp(inqb - i\omega t),$$

where q is the wave vector and ω is the frequency of the wave, nb is the equilibrium position of the n-th particle in the chain and $j = 1, 2$ denoted upper and lower layer. The Eq. (4.16) then become

$$-\omega^2 x_1 - i\beta\omega x_1 = \frac{k_b}{m_d} x_1 \left(e^{iqb} - 2 + e^{-iqb}\right) + \frac{k_d}{m_d}(x_1 - x_2) \qquad (4.18)$$

$$-\omega^2 x_2 - i\beta\omega x_2 = \frac{k_b}{m_d} x_2 \left(e^{iqb} - 2 + e^{-iqb}\right) + \frac{k_d}{m_d}(x_2 - x_1)$$
$$- \frac{k_{f0}}{m_d}(x_2 - x_1) - \frac{k_{f1}}{m_d}\left(e^{-iqb}x_1 + e^{iqb}x_1 - 2x_2\right). \qquad (4.19)$$

From trigonometric relations it is found that

$$e^{iqb} - 2 + e^{-iqb} = 2\cos qb - 2 = 2(\cos qb - 1) = -4\sin^2\left(\frac{qb}{2}\right).$$

Introducing the notations

$$\omega_b^2 = \frac{Z_d^2 e^2}{\pi \varepsilon_0 m_d b^3} = \frac{4k_b}{m_d}$$

$$\omega_d^2 = \frac{Z_d^2 e^2}{4\pi \varepsilon_0 m_d d^3} = \frac{k_d}{m_d}$$

$$\omega_{f0}^2 = \frac{Z_d Z_f e^2}{4\pi \varepsilon_0 m_d d_f^3} = \frac{k_{f0}}{m_d}$$

$$\omega_{f1}^2 = \frac{Z_d Z_f e^2}{4\pi \varepsilon_0 m_d (b^2 + d_f^2)^{3/2}} \left(\frac{3b^2}{b^2 + d_f^2} - 1 \right) = \frac{k_{f1}}{m_d}$$

where $\omega_{b,d}$ are—apart from some multiplicative constants—the dust plasma frequency which is more rigorously defined in Sect. 6.1 and ω_{f0} is the strength of the attraction. Since $Z_f \approx 0.5 Z_d$ and $d_f \approx 0.4 d$ the relative strength is $\omega_{f0}^2 / \omega_d^2 \approx 8$ which indicates that the attraction by the ion focus is much stronger than the repulsion by the upper particle. Finally, ω_{f1} is the frequency of the interactions of the lower-layer particle with the neighboring ion foci.

The equation of motion for the upper and lower layer particle then reads

$$\left[\omega^2 + i\beta\omega - \omega_b^2 \sin^2\left(\frac{qb}{2} \right) + \omega_d^2 \right] x_1 = \omega_d^2 x_2 \qquad (4.20)$$

$$\left[\omega^2 + i\beta\omega - \omega_b^2 \sin^2\left(\frac{qb}{2} \right) + \omega_d^2 - \omega_{f0}^2 + 2\omega_{f1}^2 \right] x_2 = \left(\omega_d^2 - \omega_{f0}^2 + 2\omega_{f1}^2 \cos qb \right) x_1$$

This is now the form of the equations describing the interaction between the two layers used for the stability analysis. The non-reciprocal attraction can be easily identified from the terms containing the parameters ω_{f0} and ω_{f1}. The non-reciprocity becomes obvious from the fact that these terms only appears in the equation for the lower particle, but not for the upper.

This set of equations is of fourth order in ω, but can be solved analytically by multiplying the two equations yielding

$$\lambda^2 - \lambda \left(\tilde{\omega}_1^2 + \tilde{\omega}_2^2 \right) + \tilde{\omega}_1^2 \tilde{\omega}_2^2 = \omega_d^2 (\omega_d^2 - \omega_{f0}^2 + 2\omega_{f1}^2 \cos qb), \qquad (4.21)$$

where we have used the following abbreviations

$$\lambda = \omega^2 + i\beta\omega$$

$$\tilde{\omega}_1^2 = \omega_b^2 \sin^2\left(\frac{qb}{2} \right) - \omega_d^2$$

$$\tilde{\omega}_2^2 = \omega_b^2 \sin^2\left(\frac{qb}{2} \right) - \omega_d^2 + \omega_{f0}^2 - 2\omega_{f1}^2$$

Thus, the fourth-order equation actually is bi-quadratic. The solution is

$$\omega^2 + i\beta\omega = \lambda \frac{\tilde{\omega}_1^2 + \tilde{\omega}_2^2}{2} \pm \sqrt{\frac{(\omega_{f0}^2 - 2\omega_{f1}^2)^2}{4} + \omega_d^4 - \omega_d^2(\omega_{f0}^2 - 2\omega_{f1}^2 \cos q b)}.$$

$$(4.22)$$

4.5.2 Oscillatory Instability

Here, we should step back and have a look at this solution (4.22). We do not want to stick to the exact details of the equation, but to its general form

$$\lambda = D \pm \sqrt{E^2 - F}.$$

There are two possibilities, namely that the expression under the square root is either positive or negative.

For $E^2 > F$, λ is a real number $\lambda = \lambda_r$. Then, we have

$$\omega^2 + i\beta\omega - \lambda_r = 0$$

and we would get the complex solution

$$\omega = -i\frac{\beta}{2} \pm \sqrt{\lambda_r - \frac{\beta^2}{4}}.$$

This means that the solution for $x = x_j \exp(inqb - i\omega t)$ is a damped harmonic oscillation (for $\lambda_r > \beta^2/4$). There is not much interesting about that. It does not lead to any instability. In this case, the vertical alignment is stable, both particles return with damped oscillations to their aligned state $x_{1,2} = 0$. For $\lambda_r < \beta^2/4$ we would get an exponential solution, so that the vertical alignment is absolutely unstable and the particles would move away from their aligned position. This is also a quite standard solution.

However, for the case that $E^2 < F$, λ is a complex number $\lambda_c = A \pm iB$. When $\lambda_c = A \pm iB$ is complex, the situation of stability is dramatically different. We get the same formal solution as above, namely,

$$\omega = -i\frac{\beta}{2} \pm \sqrt{A - \frac{\beta^2}{4} \pm iB}, \qquad (4.23)$$

but the imaginary contribution below the square root changes the solution drastically. There are four roots to this equation. Splitting ω into real and imaginary parts, i.e. $\omega = \omega_r + i\omega_i$, we get

$$\omega_i = -\frac{\beta}{2} \pm \frac{B}{2\omega_r} \tag{4.24}$$

$$\omega_r^2 = A + \omega_i^2 + \beta\omega_i. \tag{4.25}$$

The solution for the displacement $x = x_j \exp(inqb - i\omega_r t) \exp(-\omega_i t)$ oscillates in time with the frequency ω_r with an amplitude $x_j \exp(-\omega_i t)$ that is exponentially decaying or exponentially growing depending on the sign of ω_i. Since both signs are possible for the second term on the RHS of Eq. (4.24) positive solutions for ω_i can emerge. The change from negative to positive imaginary part ω_i can be achieved by changing the value of β.

A solution with positive imaginary part of $\omega = \omega_r + i\omega_i$ corresponds to an oscillatory instability with (exponentially) growing oscillations. Thus, such a solution would describe an instability of the vertically aligned pair of particles. The threshold value of the friction coefficient β^* for the onset of instability is found for the condition that the imaginary part is exactly zero, i.e. $\omega_i = 0$. We then have the result that

$$\omega_r = \sqrt{A} \quad \text{and} \quad \beta^* = \frac{B}{\sqrt{A}}. \tag{4.26}$$

The four roots of Eq. (4.23) are shown in Fig. 4.8 for assumed (but realistic) values of $A = 4400\,\text{s}^{-2}$ and $B = 1600\,\text{s}^{-2}$. Two of the roots always have a negative imaginary part, even for $\beta = 0$. These would correspond to damped oscillations around the vertically aligned position. The two other roots cross the $\omega_i = 0$-line at the finite friction constant $\beta^* \approx 24\,\text{s}^{-1}$. These correspond to oscillations with growing amplitudes rather than damped amplitudes. They thus are unstable oscillations. These unstable solutions are found in the entire range of the friction constant $0 < \beta < \beta^*$. This means that unstable oscillations occur *even though there is still friction of the dust particles with the background gas*. Hence, from the stability analysis we would expect that below a certain gas pressure (that controls β) unstable oscillations of the dust particles around their vertical aligned positions occur.

Thus energy for these oscillations against the neutral gas background has to be constantly supplied to the oscillation of the vertically aligned dust system. We will identify the energy source below.

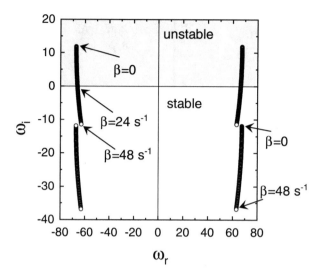

Fig. 4.8 Roots of the instability Eq. (4.23) for $A = 4400 \text{ s}^{-2}$ and $B = 1600 \text{ s}^{-2}$. When the imaginary part of the solution is larger than zero unstable oscillations with growing amplitude occur. Note, that this threshold is reached for finite values of the damping constant $\beta^* = B/\sqrt{A} = 24 \text{ s}^{-1}$

4.5.3   Conditions for the Instability

Now, we have to identify under which conditions a complex $\lambda = A \pm iB$ is obtained in Eq. (4.22) which leads to the unstable oscillations. To obtain a complex λ the square root in Eq. (4.22) must be negative. A necessary (but not sufficient) condition obviously is that

$$\omega_{f0}^2 > \omega_d^2.$$

This means that the attractive force due to the ion focus (which is represented by the parameter ω_{f0}) has to be larger than the repulsion by the upper particle (ω_d). This is expected since without domination of attraction a vertical alignment would not be found in the first place.

From a more detailed analysis (see Fig. 4.9) one finds the slightly more stringent condition that ω_{f0}^2 has to be within a certain range around $\omega_{f0}^2 \approx 2\,\omega_d^2$. Moreover, one needs $\omega_{f1}^2 > 0$ which means that the interaction with the ion foci of the neighboring particles is essential and cannot be neglected. Finally, it is seen that the instability is stronger for smaller values of $\cos qb$ indicating that mainly short wavelength modes near $qb \approx \pi$ are excited. The instability range can also be converted into a range for the ion focus charge Z_f and ion focus distance d_f as indicated for a specific situation by the dashed line. In any case, the instability manifests over a wide range of parameters.

Fig. 4.9 The shaded region in the $\omega_{f0}^2 - \omega_{f1}^2$ plane shows the range where the oscillatory instability sets in for $\cos qb = 0.1$. For $\cos qb = 0.9$ the inner, narrower range is found. Using a charge of the ion focus of $Z_f = 0.5\,Z_d$, the ion focus distance can vary between $0.46b$ and $0.77b$ to excite the instability (dashed line). Here, $b = d$ has been used

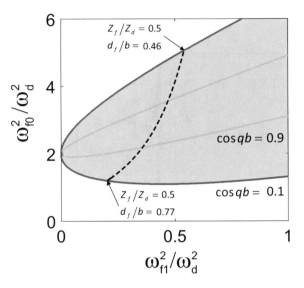

The model presented here is somewhat simplified, but it retains the main mechanisms. More sophisticated models exist [12], but in order not to focus on the mathematical details a "bare-bone" model is discussed here that allows to grasp the main ingredients of the instability.

From the physics point of view, this instability can be explained as follows: The streaming ions act as a source of free energy. The energy of the beam is transferred to the plasma crystal due to the Coulomb collisions of the ions with the dust. Above the critical value of damping the energy transferred by the ions can be dissipated by the friction of the dust particles with the neutral gas. Below the threshold the frictional losses cannot compensate the energy input and the ion energy piles up to drive the oscillatory instability.

On the level of the individual dust particles it looks like this: the lower layer particle feels an attractive force due to the ion wakefield of the upper particle. Hence it moves towards a position where it is aligned with the upper particle. However, the attractive force is non-reciprocal which means that the upper particles does *not* feel the attraction. In contrast, the upper particle feels the (screened) Coulomb repulsion from the negative charge of the lower layer particle. Hence, the upper particle tries to move away from the lower one. But the lower tries to follow the upper. The lower wants to be close to the upper, but the upper wants to escape. Due to the neighboring particles in the same layer, this type of unstable situation turns into an oscillatory instability. The neighbors of the upper particle provides a "cage" for the upper particle where it can basically oscillate at the dust plasma frequency. The lower particle tries to follow the oscillations of the upper, the upper tries to escape, and so, the oscillatory instability develops.

Summarizing, the linear stability analysis shows that a critical value of the friction coefficient $\beta^* > 0$, i.e. the gas pressure in the discharge, exists below which both upper and lower layer perform horizontal oscillations about the vertically aligned equilibrium position with exponentially increasing oscillation amplitude. These are short wavelength modes on the linear chain with $qb > 1$ with a frequency near the dust plasma frequency ω_{pd}. Moreover, from the stability analysis the relative phase and oscillation amplitude between particles of upper and lower layer can be derived. Above the critical pressure, the vertical alignment is found to be stable. This instability is directly connected to the frictional damping of the dust particles and not to a change of discharge parameters with changing pressure.

This theoretical description agrees very well with experimentally observed oscillations in two-layer plasma crystals near the melting transition [14, 28], see Fig. 4.10. The calculated values for oscillation frequency, relative phase and amplitude as well as the pressure threshold for the onset of these oscillations are within a factor of two of the measured ones which gives this model a high credibility.

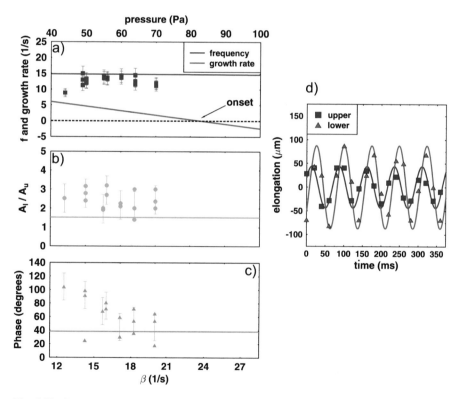

Fig. 4.10 Comparison of experimental and theoretical oscillatory instability. (**a**) Frequency of particle oscillations and growth rate of the instability, (**b**) ratio of oscillation amplitudes A_l/A_u and (**c**) phase difference of lower and upper particle. Symbols denote experimental values from oscillations during the melting transition and the lines correspond to the theoretical values from the stability analysis. (**d**) One experimental example of horizontal oscillations of a vertically aligned pair. After [28]

We will return to these oscillations when dealing with the phase transition of the plasma crystal in the following chapter.

References

1. U. Konopka, G. Morfill, L. Ratke, Phys. Rev. Lett. **84**, 891 (2000)
2. J.H. Chu, I. Lin, Phys. Rev. Lett. **72**, 4009 (1994)
3. T. Trottenberg, A. Melzer, A. Piel, Plasma Sources Sci. Technol. **4**, 450 (1995)
4. A. Melzer, A. Homann, A. Piel, Phys. Rev. E **53**, 2757 (1996)
5. U. Mohideen, H. Rahman, M.A. Smith, M. Rosenberg, D.A. Mendis, Phys. Rev. Lett. **81**, 349 (1998)
6. B.A. Annaratone, A.G. Khrapak, A.V. Ivlev, G. Söllner, P. Bryant, R. Sütterlin, U. Konopka, K. Yoshino, M. Zuzic, H.M. Thomas, G.E. Morfill, Phys. Rev. E **63**, 036406 (2001)
7. S.V. Vladimirov, M. Nambu, Phys. Rev. E **52**, R2172 (1995)
8. M. Lampe, G. Joyce, G. Ganguli, Phys. Plasmas **7**, 3851 (2000)
9. M. Nambu, S.V. Vladimirov, P.K. Shukla, Phys. Lett. A **203**, 40 (1995)
10. P. Ludwig, W.J. Miloch, H. Kählert, M. Bonitz, New J. Phys. **14**, 053016 (2012)
11. O. Arp, J. Goree, A. Piel, Phys. Rev. E **85**, 046409 (2012)
12. V.A. Schweigert, I.V. Schweigert, A. Melzer, A. Homann, A. Piel, Phys. Rev. E **54**, 4155 (1996)
13. V.A. Schweigert, V.M. Bedanov, I.V. Schweigert, A. Melzer, A. Homann, A. Piel, JETP **88**, 482 (1999)
14. A. Melzer, V.A. Schweigert, I.V. Schweigert, A. Homann, S. Peters, A. Piel, Phys. Rev. E **54**, R46 (1996)
15. V.R. Ikkurthi, K. Matyash, A. Melzer, R. Schneider, Phys. Plasmas **16**, 043703 (2009)
16. W.J. Miloch, M. Kroll, D. Block, Phys. Plasmas **17**, 103703 (2010)
17. A. Piel, Phys. Plasmas **18**, 073704 (2011)
18. I.H. Hutchinson, Phys. Rev. Lett. **107**, 095001 (2011)
19. J. Schleede, L. Lewerentz, F.X. Bronold, R. Schneider, H. Fehske, Phys. Plasmas **25**, 043702 (2018). https://doi.org/10.1063/1.5021316
20. A.V. Ivlev, J. Bartnick, M. Heinen, C.R. Du, V. Nosenko, H. Löwen, Phys. Rev. X **5**, 011035 (2015). https://link.aps.org/doi/10.1103/PhysRevX.5.011035
21. A. Piel, H. Jung, F. Greiner, Phys. Plasmas **25**, 083703 (2018). https://doi.org/10.1063/1.5039606
22. K. Takahashi, T. Oishi, K. Shimomai, Y. Hayashi, S. Nishino, Phys. Rev. E **58**, 7805 (1998)
23. A. Melzer, V. Schweigert, A. Piel, Phys. Rev. Lett. **83**, 3194 (1999)
24. A. Melzer, V. Schweigert, A. Piel, Phys. Scr. **61**, 494 (2000)
25. G.A. Hebner, M.E. Riley, Phys. Rev. E **68**, 046401 (2003)
26. M. Kroll, J. Schablinski, D. Block, A. Piel, Phys. Plasmas **17**, 013702 (2010)
27. L. Couëdel, V. Nosenko, A.V. Ivlev, S.K. Zhdanov, H.M. Thomas, G.E. Morfill, Phys. Rev. Lett. **104**, 195001 (2010)
28. A. Homann, *Schmelzprozeß und Dynamik des Plasmakristalls*. Dissertation (Christian–Albrechts–Universität Kiel, Kiel, 1998)

Chapter 5
Plasma Crystallization and Phase Transitions

After identifying the basic mechanisms of charging, trapping and interaction of dust particles in a discharge, we now investigate the many-particle interaction of the dust in view of crystallization and phase transitions of the dust ensemble. One of the fascinating properties of dusty plasmas is that the dust particles can arrange in highly ordered systems and that a transition to unordered situations is found. Thus, dusty plasmas can serve as models for condensed matter systems. Now, here, it is discussed how the ordering of a dust particle ensemble can be characterized, especially for situations in two dimensions.

5.1 The One-Component Plasma (OCP)

At first, we like to investigate how systems of many charged particles behave under their mutual electrostatic influence. The simplest situation that can be considered is that of point charges immersed in a homogeneous neutralizing background. As early as in the 1930s, Eugene Wigner [1] has discussed this situation in the context of (positive) ions immersed in a "sea" of electrons as a model to study the crystalline order in metals. In dusty plasmas, we obviously identify the point charges with the (negative) dust particles and the neutralizing background with the (positive) plasma ions. In such a system with a homogeneous background the point charges interact by pure Coulomb repulsion. The background is only necessary to ensure overall charge neutrality. Such a system is known as a one-component-plasma (OCP).

Such a system of pure point charges is characterized by only a single parameter, the Coulomb coupling parameter

$$\Gamma = \frac{Q^2}{4\pi \varepsilon_0 b_{\mathrm{WS}}} \frac{1}{k_{\mathrm{B}} T} . \tag{5.1}$$

© Springer Nature Switzerland AG 2019
A. Melzer, *Physics of Dusty Plasmas*, Lecture Notes in Physics 962,
https://doi.org/10.1007/978-3-030-20260-6_5

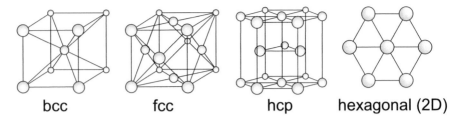

bcc   fcc   hcp   hexagonal (2D)

Fig. 5.1 Crystal structures in 3D: BCC, FCC and HCP. In 2D: hexagonal crystal structure. The BCC structure contains particles at the edges of a cube and an additional particle in the cube center. Likewise FCC contains additional particles at the cube faces. HCP is very similar to FCC where the ordering repeats every other layer instead of every third layer as in FCC

It describes the electrostatic interaction of neighboring point charges in relation to their thermal energy. Here, T is the temperature of the point charge species and Q is their charge. As a measure of the interparticle distance the Wigner-Seitz radius $b_{WS} = (4\pi n/3)^{-1/3}$ is defined with n being the number density of the point charges[1].

A charged particle system is said to be strongly coupled when $\Gamma > 1$, i.e. when the electrostatic energy of neighboring particles exceeds the thermal energy. A usual electron-ion plasma Γ is weakly coupled with Γ being much smaller than unity.[2]

From simulations of charged-particle systems it was found that the point charges arrange in ordered crystalline arrangements when the coupling parameter exceeds a critical value of $\Gamma_c = 168 \pm 2$ [2]. For $\Gamma < \Gamma_c$ the particles are in a fluid (= liquid or gas-like) state. Hence, there is a solid-liquid phase transition at $\Gamma_c = 168$. Note, that a purely repulsive system has only a solid-fluid transition, there is no liquid-gas transition. A liquid-gas transition requires an attractive part in the interaction potential. So, when the Coulomb energy by far dominates the thermal agitation the particles crystallize into an ordered arrangement, a process known as **Wigner** or **Coulomb crystallization**. In 3D, the particles arrange in a BCC lattice[3] (see Fig. 5.1) for the different lattice types) which is the minimum energy configuration for pure Coulomb interaction.

As known from solid-state textbooks [3], the minimum energy configuration is determined from the sum of the electrostatic energies between all particles, the so-

---

[1] For example, in an ordered state with simple cubic structure with particle separation b the density is $n = b^{-3}$. The Wigner-Seitz radius for this situation then is $b_{WS} = 0.62\,b$.

[2] For example, one finds $\Gamma = 8 \times 10^{-3} \ll 1$ for ions at $T_i = 0.03\,\text{eV}$ and $n_i = 1 \times 10^9\,\text{cm}^{-3}$ and even less for electrons due to their (typically) higher temperature.

[3] BCC: body-centered cubic, FCC: face-centered cubic, HCP: hexagonal-close packed.

called Madelung energy,

$$U = \frac{1}{2N} \sum_{i \neq j}^{N} \frac{Q^2}{4\pi \varepsilon_0 r_{ij}}, \tag{5.2}$$

where r_{ij} is the relative distance between particle i and j. From detailed calculations the Madelung energy is found for pure Coulomb interaction as [4]

$$U_{\text{BCC}} = -0.895\,929\,255\,682 \text{ for the BCC structure}$$

$$U_{\text{FCC}} = -0.895\,873\,615\,195 \text{ for the FCC structure}$$

$$U_{\text{HCP}} = -0.895\,838\,120\,459 \text{ for the HCP structure}$$

in units of $\Gamma = Q^2/(4\pi \varepsilon_0 b_{\text{WS}})$. The energies for the different lattice types are very close to each other, but the BCC structure is the one with the lowest energy.

In 2D systems crystallization takes place at the critical value $\Gamma_c = 125$ [2]. In 2D, the Wigner-Seitz radius is defined as $b_{\text{WS}} = (\pi n)^{-1/2}$. The minimum energy configuration is the hexagonal structure as shown in Fig. 5.1. This is the structure you would expect from densely covering a table with coins.

5.2 Yukawa Systems

When shielding of the point charges is taken into account the neutralizing background is no longer homogeneous. As a consequence, the interaction potential between the point charges is described by the Debye-Hückel (Yukawa) potential, see Eq. (4.1). As we have seen in Sect. 4.1 this is a good approximation for a dust system, at least in the absence of wake field effects.

Besides the Coulomb coupling parameter, Yukawa systems are characterized by a second parameter, the so-called screening strength,

$$\kappa = \frac{b_{\text{WS}}}{\lambda_D} \tag{5.3}$$

which is the Wigner-Seitz distance b_{WS} in units of the Debye length. The OCP-limit of $\Gamma_c = 168$ is retrieved for infinite screening length, i.e. for $\kappa \to 0$.

The phase diagram of the Yukawa system is shown in Fig. 5.2. First, it is seen that the critical value for the solid-fluid transition strongly depends on the screening strength, i.e. $\Gamma_c = \Gamma_c(\kappa)$. The melting line increases almost exponentially with κ. Due to the exponential shielding a much higher Coulomb coupling parameter is required to enter the crystalline regime [5, 6].

Fig. 5.2 Phase diagram of the Yukawa system in the Γ-κ plane. The melting line increases almost exponentially with the screening strength. In the solid phase two different crystal structures, BCC and FCC are found. Redrawn after the data of [6]

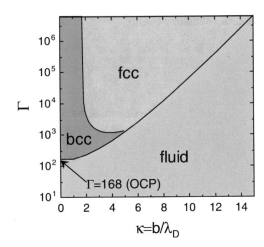

The solid phase itself shows two different crystalline structures: for lower values of κ the BCC structure as in the OCP is found. For stronger screening the FCC structure is obtained. For increasing screening strength the Yukawa interaction becomes more and more like a hard-sphere potential. The packing density for spheres in the BCC structure is 68% whereas it reaches 74% in the FCC structure (the FCC structure is what one would obtain from stacking spheres, e.g. oranges, into several layers). Thus at higher screening an increased packing density becomes more favorable.

5.3 Coulomb Crystallization in Dusty Plasmas

After this general introduction to strongly coupled systems we now investigate under which conditions Coulomb crystallization in dusty plasmas will be possible. The following ideas have already been presented in 1986 by Ikezi [7], years before plasma crystals have been discovered experimentally. Ikezi's model uses a few simplifying assumptions which however capture the main features of the problem. First, the melting line of Yukawa systems is approximated by an effective Coulomb coupling parameter

$$\Gamma_{c,\text{eff}} = \frac{Z_d^2 e^2}{4\pi\varepsilon_0 b_{\text{WS}} k_B T_d} \exp\left(-\frac{b_{\text{WS}}}{\lambda_D}\right) = \Gamma \exp(-\kappa) = 168\,,$$

where it is assumed that the melting line exactly increases exponentially with increasing κ. Above $\Gamma_{c,\text{eff}} = 168$ crystallization is achieved in Ikezi's model. Since $\lambda_D \approx \lambda_{D,i}$ from Eq. (4.2) only ion screening is considered. The second assumption is concerned with the charging of the dust. For the dust charging the rule-of-thumb expression according to Eq. (2.22) is used. The dust charge, however, is limited

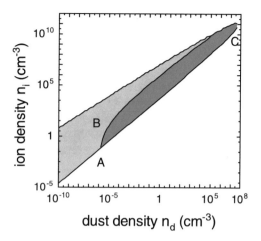

Fig. 5.3 Existence diagram of Wigner crystals in dusty plasmas. In the dark area Coulomb crystallization should be possible for a particle of $10\,\mu m$ radius. In the total shaded area crystallization can occur for arbitrarily sized particles. Note the wide logarithmic scale on both axes. After [7]

when all electrons are bound to the dust (electron depletion, see Sect. 2.6), i.e.

$$Z_{d,\lim} = \frac{n_i}{n_d}.$$

Now, under which parameters of ion and dust density, Wigner crystallization in dusty plasmas can occur? The ion density affects the screening length λ_D and the dust charge limit. High ion density means high dust charge limit, but strong screening. The dust density also affects the dust charge limit and the interparticle distance b. High dust densities mean low maximum dust charge, but small interparticle distance which results in high coupling. Both parameters thus have counteracting effects. From these considerations, Ikezi found that Wigner crystallization should be possible in the dark area ABC in Fig. 5.3. The boundary of this region is dominated by different mechanisms. From A to B, the dust density is roughly constant. Also the charge Z_d is constant, there are no depletion effects due to the low dust density. On the entire upper boundary from A to C via B the particle is charged to its single particle value (here $a = 10\,\mu m$ is assumed). From A to B, the Debye length is much larger than the interparticle distance and screening effects are not important. Near B the situation changes. The ion density becomes so high that the Debye length is now of the order of the interparticle distance and screening becomes dominant. Thus the boundary bends towards much higher dust densities and thus smaller interparticle distances until point C is reached. On the boundary from C to A there are relatively large dust densities and low ion densities. Here, depletion effects become dominant. The charge on the dust is determined by the available free electrons thus limiting the coupling parameter.

Although this is a quite crude model it shows that Coulomb crystallization in dusty plasmas is possible in a range of ion and dust densities that is several orders of magnitude wide. For typical plasma discharges with $n_i = 10^9 – 10^{10}\,cm^{-3}$ plasma crystals should exist for dust densities in the range from $n_d = 10^3 – 10^5\,cm^{-3}$.

5.4   Crystallization in Bounded Systems

In the previous sections, the formation of Wigner crystals in extended 3D (and 2D) systems has been discussed. However, as we have seen, the plasma crystals in the experiments (see Fig. 3.8) are extended in the horizontal plane, but are confined in the plasma sheath by quite strong vertical fields. Similar confined systems are also found in experiments on ions trapped in quadrupole traps [8, 9].

The situation of confined OCP systems has to capture features of both 2D and 3D systems with competing planar hexagonal or volume BCC/FCC structures. To study possible crystal structures in bounded systems, Dubin has considered a system of point charges which is extended in the horizontal $x-y$ plane, but is confined by a parabolic potential in the vertical z-direction [10], see Fig. 5.4. When the number of point charges is not too high the particles will arrange in a monolayer 2D crystal with hexagonal structure as discussed above. When the density of charges is increased (or, equivalently, the vertical confinement is weakened) the monolayer system jumps to two layers, three layers and so on. This is an expected behavior of repulsive particles. It is interesting to note here that the crystal structure of the multi-layer system changes between square, BCC 110 and hexagonal. Thus the possible crystal structures of the infinite 3D systems are also found in a system with only a few layers.

Fig. 5.4 Layer formation and crystalline structure in a confined system: Vertical position of the individual crystal layers. The respective crystal structure is indicated by the three inserts at the top, where the structure is viewed from top. Different shades of the circles denote different layers. Reprinted figure with permission from [10]. ©(1993) by the American Physical Society

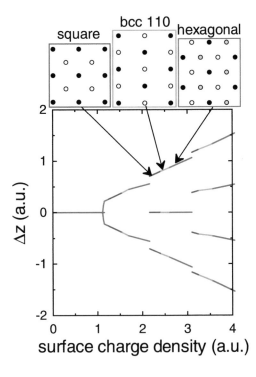

The scenario is not much different if screening between the charges is taken into account [11, 12]. There, similar jumps to multi-layer systems with similar crystal structures are observed. However, the exact transition points, of course, depend on the screening strength. Structures of systems confined in all spatial dimensions are described in Chap. 8.

5.5   Structural Information

To characterize the degree of order of the point charges and to define a system as fluid or solid it is necessary to have quantitative measures. Here, such techniques are presented with special emphasis on 2D systems, but they can be extended to 3D systems. We will exploit these findings in the discussion of the phase transitions below.

Wigner-Seitz Cells   One of the basic techniques to characterize a system of point charges is given by the Wigner-Seitz cell analysis. There, the 2D plane is covered with polygons where each particle is in the center of its own cell. The construction is as follows: for each particle, the perpendicular bisector of the connection to neighboring particles is determined. The bisectors define a closed polygon around the particle of interest (see Fig. 5.5a). The main advantage of this procedure is that it determines which particles in the vicinity of the reference particle are the neighboring particles. Thus, the number of nearest neighbors is also determined. For a 2D system, the *mean* number of neighbors is 6. However, particles with 5 or 7 neighbors might exist which are termed defects. The relative abundance of 6-fold polygons is already a good qualitative hint whether a structure is crystalline or not. A structure with more than 90% of hexagons can be called crystalline (see Fig. 5.7).

Pair Correlation Function   The pair correlation function $g(r)$ is the probability to find a particle at a distance r from a chosen particle. This is then averaged over all

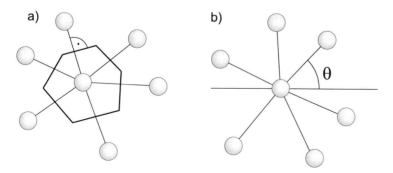

Fig. 5.5 (a) Sketch of the Wigner-Seitz cell construction. (b) Definition of the bond angle θ

Fig. 5.6 Pair correlation function $g(r)$ for various coupling parameters Γ of the fluid OCP system ranging from weakly coupled ($\Gamma = 1$) to almost crystalline ($\Gamma = 140$). The nearest neighbor distance is found to be at $r \approx 1.7 b_{WS}$. After [2, 13]

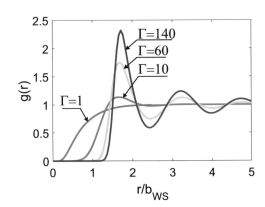

(chosen) particles. One finds that

$$g(r) \rightarrow 1 \qquad \text{for } r \rightarrow \infty,$$

since one will always find a particle at large distances, and

$$g(r) \rightarrow 0 \qquad \text{for } r \rightarrow 0,$$

since the particles cannot come infinitely close to each other.

The pair correlation function gives the mean interparticle distance as well as the near and far order of the particle system. An example is shown in Fig. 5.6 for simulated OCP systems at different coupling strengths. The first peak in $g(r)$ is due to the neighboring particles, further peaks due to overnext neighbors and so on. For a 2D hexagonal structure one would expect a peak at the interparticle distance b, the next peaks would be at $\sqrt{3}b$ and $2b$.

The more pronounced the peaks are and the more peaks can be identified the higher is the order of the system which also can be seen from Fig. 5.6. For small values of the coupling parameter ($\Gamma \approx 10$) there is only a very small hump at the nearest neighbor distance and the pair correlation function is nearly flat. This means that there is no strong order in the system and one can find particles at all distances larger than the interparticle distance. When Γ is increased the nearest neighbor peak grows and also the peaks of overnext neighbors (and so on) grow, reflecting increased order of the system. This can also be easily seen in the comparison of solid, fluid and gas-like states in Fig. 5.7.

One might expect that at very high coupling parameters $\Gamma \gg \Gamma_c$ the particles should be at the exact lattice sites and the pair correlation function should collapse to single sharp peaks. However it was derived from very basic principles [14, 15] that there are always long-range fluctuations in 2D systems that destroy order over very long distances. It has also been argued whether a phase transition in 2D might be a two-step phase transition of 2nd order via an intermediate so-called hexatic phase [15]. For such a phase transitions in 2D systems the form of $g(r)$ would

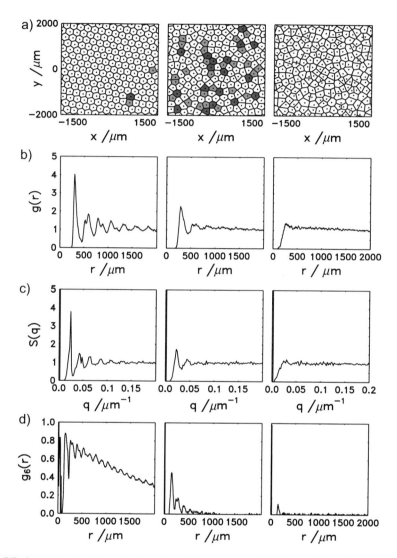

Fig. 5.7 Structure of solid and fluid and gas-like structures, (**a**) Wigner-Seitz cell construction. For the solid and liquid state the defects are shaded (blue: 5-fold defect, red: 7-fold defect). (**b**) Pair correlation function $g(r)$, (**c**) structure factor $S(q)$ and (**d**) angular correlation function $g_6(r)$

exhibit a distinct long-range decay. Also the behavior of defects would follow a certain scenario. Some of these aspects have been studied for dust systems [16–18], but that will not be further explicated here.

Structure Factor In condensed matter physics, the pair correlation function usually is not determined directly. Rather, the structure factor is determined from scattering of x-rays or neutrons. In dusty plasmas, we are able to measure the pair correlation function directly and to *calculate* the structure factor in order to compare with condensed matter experiments. The structure factor $S(\mathbf{q})$ essentially is just the Fourier transform of the pair correlation function

$$S(\mathbf{q}) = 1 + n_d \int (g(r) - 1) \exp(i\mathbf{q} \cdot \mathbf{r}) \, d\mathbf{r}, \tag{5.4}$$

where $\mathbf{q}$ is the wave vector of the scattered radiation. In 2D, this can be written as

$$S(q) = \frac{1}{N} \sum_{ij}^{N} \frac{1}{2\pi} \int_0^{2\pi} \exp(i q r_{ij} \cos\varphi) \, d\varphi = \frac{1}{N} \sum_{ij}^{N} J_0(q r_{ij}),$$

where φ is the angle between $\mathbf{r}$ and $\mathbf{q}$ and J_0 is the zero-order Bessel function. The calculated structure factor is also shown in Fig. 5.7.

A large first peak in the structure factor means long-range periodicity of the pair correlation function. Thus, a peaked structure factor corresponds to long-range order.

Angular Correlation Function Finally, the angular correlation will be discussed, here. Until now, only the translational order has been characterized. In contrast, the angular correlation function measures how the bond angles between nearest neighbors are oriented relative to each other as a function of distance between the bonds. The angular correlation is defined as

$$g_6(r) = \langle \exp(6i[\theta(r) - \theta(0)]) \rangle, \tag{5.5}$$

where θ is the angle of a nearest neighbor bond relative to a fixed axis (see Fig. 5.5b). The factor 6 takes into account the presumed hexagonal order of the system and $\langle \cdot \rangle$ denotes the average over the particle arrangement. When $g_6(r)$ is close to 1 the nearest-neighbor bonds at a relative distance r are oriented along the same direction. This means that the Wigner-Seitz cell at one place in the crystal is aligned in the same direction as the Wigner-Seitz cell a distance r apart.

The bond orientational order is destroyed by defects (particles with 5 or 7 neighbors). Behind defects the lattice orientation is different from the starting point. A value of g_6 close to 0 means that the bonds are randomly arranged (no correlation), a value of $g_6 = -1$ describes anti-correlation: the bonds have a relative orientation which differs by 30° which is the maximum difference in angle for a 6-fold symmetry. The angular correlation function of our example systems is also shown in Fig. 5.7. The angular correlation function also plays a large role in the identification of melting processes in 2D systems [15].

The bond order averaged over the nearest neighbor bonds

$$|\psi_6| = \left| \frac{1}{M} \sum_{m=1}^{M} \exp(6i\theta_{jm}) \right| \tag{5.6}$$

can be used as a local order parameter. Here, M is the number of nearest neighbors of particle j and θ_{jm} is the bond angle from particle j to its neighboring particles m [19]. Then, $|\psi_6| = 1$ at a perfect sixfold lattice site and its value decreases when the local ordered structure is distorted, see also Fig. 5.9.

Numerical Example The different techniques to characterize a particle arrangement is illustrated for three systems that have been obtained from numerical modeling. The three examples in Fig. 5.7 reflect crystalline ($\Gamma > \Gamma_c$), fluid ($\Gamma > 1$) and gas-like ($\Gamma < 1$) order (it should be noted again, here, that for a system with only repulsive forces like OCP or Yukawa there is only a solid-fluid transition). In the solid state the particles are nicely arranged, only few defects exist. Hence, the pair correlation function is very pronounced. Pair order can be observed at least up to 7 interparticle distances. Correspondingly, the structure factor is large and sharp. The angular correlation function is large and decays only slowly with distance. In the fluid state, one finds larger number of defects, but one can still identify the nearest-neighbor and overnext-neighbor peak in $g(r)$. The structure factor thus is smaller and not as sharp as in the solid state. Also the angular correlation rapidly decreases. The gas-like state exhibits no correlation at all. The pair correlation is flat (compare Fig. 5.6 at $\Gamma = 10$), the structure factor is flat and there is no angular correlation, even for the smallest distances. This reflects the gas-like characteristics of this particle arrangement.

5.6 3D Crystals

The previous discussions were related to some extent to 2D crystals (or crystals with a few layers) since they are easily produced in laboratory experiments. Nevertheless, three-dimensional dust crystals have been observed under microgravity conditions aboard the ISS [20–22] where the particles are not forced into the space charge sheath, but remain trapped in the plasma volume. Figure 5.8a shows the reconstruction of the 3D positions of some 10,000 particles forming an ordered arrangement.

The identification of local order is much more difficult in 3D than in 2D due to the much more possibilities of particle arrangements. In a particular set of experiments [20, 21], it was tried to identify the local crystal structure. For that purpose, for each particle of the crystal a local order parameter q_4 and q_6 (the 3D equivalent to $|\psi_6|$) was calculated that accounts for local 4- and 6-fold order using the order properties of spherical harmonics [20, 21, 23]. The abundance of particles in the crystal with local order parameters q_4, q_6 is given in Fig. 5.8b where also the order parameters of ideal HCP and FCC lattices (and also BCC which is not shown here) are indicated (compare also Fig. 5.1). It is seen that quite a substantial fraction of the particles are located close to the ideal HCP and FCC order parameters indicating high crystallinity (either as HCP or FCC lattice). However, also a certain fraction of the particles are randomly oriented showing that the system is not fully crystallized.

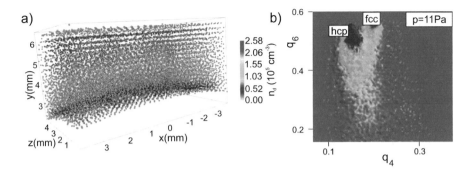

Fig. 5.8 (**a**) 3D dust crystal under microgravity conditions aboard the ISS. (**b**) Color-coded number of particles with local order parameters q_4, q_6 indicating local 4- and 6-fold symmetry in another experiment. Reprinted figures with permission from [21, 22]. © (2012, 2016) by the American Physical Society

In the plasma volume the forces acting on the particles are much smaller than in the sheath. Hence also the interparticle forces are usually much smaller since they have to compete with much smaller confining forces. Hence, such 3D crystals are usually much softer than those in the sheath (although those are already very soft). Hence, crystallization occurs on a very, very long time scale. Hence, formation (and observation) of 3D crystals is an interesting and difficult problem.

5.7 The Plasma Crystal

We now turn to experimental realizations of nearly 2D ordered plasma crystals and their phase transition.

Due to the high charges on the dust particles a dust system usually is far in the strong-coupling range with large Coulomb coupling parameters Γ [24–26]. In the experiment, dust particles often arrange in an ordered state, see Fig. 1.3. There, at high gas pressures (118 Pa in this particular experiment) a well ordered crystalline structure is found as can be seen in Fig. 5.9 from the pair correlation $g(r)$ and angular correlation function $g_6(r)$ for this crystal, compare also with Fig. 5.7. From $g(r)$ the interparticle distance is obtained as $b = 450\,\mu\text{m}$. The local order parameter $|\psi_6|$ is generally very high in this ordered crystal. Low values of $|\psi_6|$ indicate defect sites.

Fig. 5.9 (a) pair correlation $g(r)$ and (b) angular correlation function $g_6(r)$ for the crystal in Fig. 1.3 (see also inset). In (c) the Wigner-Seitz cells for this crystal are shown. The cells have been colored according to the local bond-order parameter $|\psi_6|$

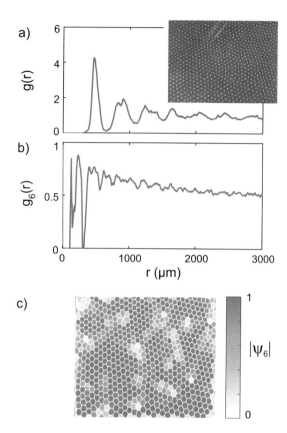

The dust charge was measured from the resonance technique as $Z_d \approx 5000$ [17]. The dust particle temperature[4] is the mean kinetic energy of the dust particles

$$\frac{1}{2}m_d\langle v_\alpha^2\rangle = \frac{1}{2}k_B T_d , \qquad (5.7)$$

where v_α is the dust velocity in a particular spatial direction $\alpha = x, y, z$, see also [30]. In this ordered case, the dust temperature is derived essentially as room temperature. The corresponding Coulomb coupling parameter then is $\Gamma \approx 2000$ which indeed indicates a crystalline state.

Further, when looking at the long-time motion of the dust particles, see Fig. 5.10a where the dust trajectories over 10 s are shown, the particles do not move considerably in the crystalline state, they stay in their respective Wigner-Seitz cells.

---

[4]It is not the surface temperature of the dust which due to the contact with the cold neutral gas generally is around room temperature, see also [27–29].

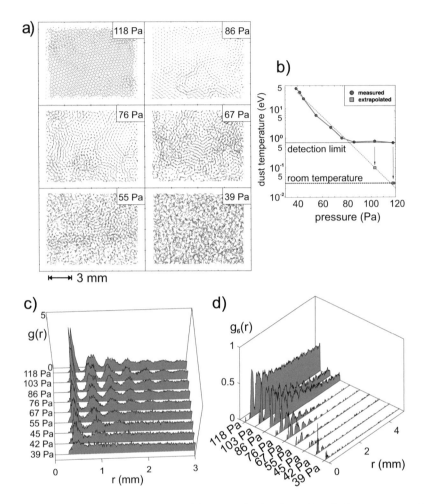

Fig. 5.10 (**a**) Trajectories of the dust particles over 10 s for decreasing discharge pressure, (**b**) temperature of the dust particles as a function of discharge pressure. A temperature below 0.7 eV could not be detected due to the limited optical resolution, (**c**) pair correlation and (**d**) orientational correlation function versus pressure. After [17]

5.8 The Phase Transition of the Plasma Crystal: Experiment

We now will study how phase transitions of plasma crystals are achieved in the experiment. The transition is due to heating of the dust component due to the oscillatory wakefield-induced instability discussed in Sect. 4.5. Hence, the experiments have been performed with two-layer plasma crystals which are subject to this instability. Mono-layer crystals do not show a melting scenario as described below.

Experimentally, the melting transition is observed when the gas pressure in the discharge is reduced [17, 31]. At high gas pressures (118 Pa in this experiment) the just discussed well ordered crystalline structure is found (Fig. 5.10a). At reduced pressure the particle arrangement undergoes a transition to a liquid and, finally, to an almost gas-like state. During this transition, at first, stream line particle motion around crystalline patches sets in, that gradually turns into a more and more irregular particle motion. This transition is also seen in the correlation function $g(r)$ and $g_6(r)$, see Fig. 5.10c, d. The translational as well as the orientational ordering is strongly decreased from the ordered state at high pressures to the completely disordered, gas-like state at 39 Pa.

The transition from the ordered to the liquid state is accompanied, and even driven, by the horizontal oscillations of the vertically aligned pairs that were discussed in the previous section. One can easily observe these oscillations around the equilibrium position by video microscopy. They are not visible in the trajectories since these are averaged over 10 s. The oscillations have already been described in Fig. 4.10.

From the thermodynamic point of view, the transition is driven by a dramatic increase of dust temperature from room temperature at 118 Pa to $T_d = 50\,\text{eV}$ at 39 Pa, see Fig. 5.10b. This high temperature is far above the energies of all other plasma species. This dramatic increase of dust temperature cannot be explained by simple changes in the discharge conditions with reduced pressure. In this pressure range of interest the densities of electrons and ions change by a factor about 3 and the electron temperature is nearly unaffected [17]. These small changes cannot be made responsible for the change of dust temperature by a factor 1000.

Naturally, the previously described instability mechanism based on the non-reciprocal attraction is the source of energy input to the microspheres. The oscillatory instability sets in at about 80 Pa under the conditions of this experiment (see Fig. 4.10). This is exactly the pressure when fluid particle trajectories set in. The energetic oscillations are becoming more and more irregular and turn into a chaotic motion of the particles which can be interpreted as heating the dust particles to the high temperatures mentioned above.

One should note that very similar phase transitions have been observed in krypton at a lower gas pressure, but at nearly the same values for the frictional damping constant β [31].

Melting transitions of single-layer crystals have also been observed [32]. The melting mechanism is, however, different from the one discussed here, but also relies on the presence of the ion focus. In a monolayer crystal with ion focus the horizontal motion of the particles can couple to the vertical motion [33–36] which again can lead to an instability. The role of this mode coupling in two-layer systems in relation to the above described oscillatory instability has been discussed in [37].

5.9   The Phase Transition of the Plasma Crystal: Simulations and Theory

To identify the heating mechanism and melting dynamics of plasma crystals the melting transition was investigated in simulations in which the experimental conditions were mimicked in great detail [38], see also [39]. Some of the interesting questions are how the instability mechanism drives the transition and how the oscillatory particle motion is randomized to heat the crystal.

In the simulations [38] an ideal hexagonal two-layer plasma crystal is considered. The negatively charged particles interact via a screened Coulomb repulsion and the particles of the lower layer experience the non-reciprocal, attractive force from the ion clouds of the upper particles. Here, the full horizontal dependence of the attractive force according to Fig. 4.4a is taken into account. In the simulation only horizontal displacements $\mathbf{r}_{ij} = (x_{ij}, y_{ij})$ from the equilibrium positions are taken into account ($j = 1, 2$ indicates upper and lower layer, respectively, and i denotes the number of the particle in that layer):

$$\frac{d^2\mathbf{r}_{ij}}{dt^2} + \beta\frac{d\mathbf{r}_{ij}}{dt} = \frac{1}{m_d}\mathbf{F}_{ij} + \frac{1}{m_d}\mathbf{F}_L. \tag{5.8}$$

The force acting on the particles $\mathbf{F}_{ij}$ consists of two parts (compare Fig. 4.7). The first describes the repulsion due to the other dust particles in the same or in the other layer. The second part is on the lower layer particles only, and describes the attractive force on the lower particles due to the ion focus (replaced by a single point charge). In addition, a Langevin force F_L is applied to the particles to give them a finite temperature (room temperature) in the crystalline state at high pressure. This Langevin force is kept constant at room temperature throughout the melting process and does not lead to additional heating.

This model is exactly the same as that used for the linear stability analysis in Sect. 4.5 with the exception of the Langevin force and the consideration of the full horizontal dependence, and not only the linear part for small elongations. As in the experiment, the gas pressure, i.e. the friction constant β, was slowly reduced in the simulation. The resulting energy of the dust particles as a function of gas pressure is shown in Fig. 5.11 together with the particle trajectories for different values of the friction coefficient.

For the melting transition of an ideal crystal a two-step melting scenario is obtained: Starting at high pressures the well ordered crystal is found. With decreasing friction the oscillatory instability described in Sect. 4.5 sets in at about $\beta_{in} = 0.165\omega_{pd}$. This leads to a dramatic increase of the dust kinetic energy from room temperature to about 10 eV. However, that does not lead to the melting of the crystal, instead, a hot crystalline state is found, here. With further reduction of friction ($\beta_* = 0.12\omega_{pd}$) the transition to a liquid state is observed.

The maximum energy of the dust particles and the overall melting scenario in the simulation agrees well with that obtained from experiments. However, some

Fig. 5.11 (a) Particle trajectories for different friction coefficients during the melting transition. (b) Particle energy of upper and lower layer versus friction coefficient. The points of which the trajectories are shown in (a) are indicated by arrows. The two curves correspond to the energy of upper and lower layer, respectively. From [38]

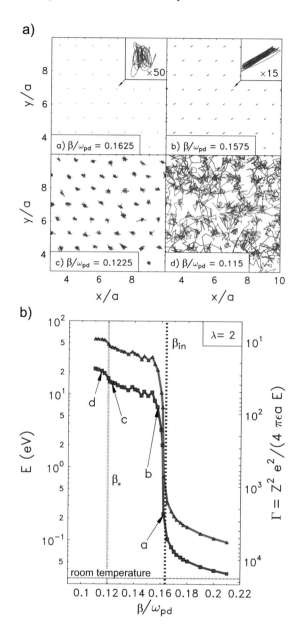

differences in comparison to the experiment are found in the simulation. First, the experimental melting transition takes place over a broader range of β than the simulation. Second, the two-step melting is not clearly observed in the experiment and, third, in the experiment characteristic stream line particle motions around crystalline patches are seen that are not present in the simulation.

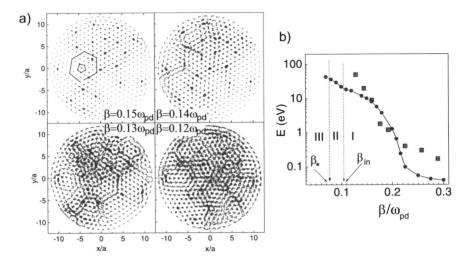

Fig. 5.12 (**a**) Particle trajectories for different friction coefficients during the melting transition. In the upper left panel the 5-fold defect is marked with a pentagon. (**b**) Particle energy versus friction coefficient. Open circles: Two-step melting of a crystal with defects (simulation), squares: Experimental results. From [40]

This discrepancy is resolved when also defects in the plasma crystal are taken into account [40]. The crystal with defects shows a melting scenario that is very similar to the experimental: the increase of dust temperature with reduced pressure is slower, in overall quantitative agreement with the experiment. In addition, the two-step melting transition is "smeared out" over a broader range of β so that the melting is gradual and no sharp transition can be assigned. Finally, the particle trajectories show the streamline particle motions that are characteristic for the experiment. The melting of a crystal with defects is shown in Fig. 5.12. Compare this simulated phase transition with the experimental one in Fig. 5.10. One can observe a deep agreement between the experimental and simulated phase transition. Since the simulations are only based on the non-reciprocal attraction as discussed above and no other heating mechanisms are taken into account this agreement gives strong confidence in the existence and relevance of these non-reciprocal forces.

References

1. E. Wigner, Trans. Faraday Soc. **34**, 678 (1938)
2. S. Ichimaru, Rev. Mod. Phys. **54**, 1017 (1982)
3. C. Kittel, *Introduction to Solid State Physics*, 6th edn. (Wiley, New York, 1986)
4. D.A. Baiko, A.Y. Potekhin, D.G. Yakovlev, Phys. Rev. E **64**, 057402 (2001). https://link.aps.org/doi/10.1103/PhysRevE.64.057402
5. M.O. Robbins, K. Kremer, G.S. Grest, J. Chem. Phys. **88**, 3286 (1988)
6. S. Hamaguchi, R. Farouki, D.H.E. Dubin, Phys. Rev. E **56**, 4671 (1997)

7. H. Ikezi, Phys. Fluids **29**, 1764 (1986)
8. I. Waki, S. Kassner, G. Birkl, H. Walther, Phys. Rev. Lett. **68**, 2007 (1992)
9. A. Mortensen, E. Nielsen, T. Matthey, M. Drewsen, Phys. Rev. Lett. **96**, 103001 (2006)
10. D. Dubin, Phys. Rev. Lett. **71**, 2753 (1993)
11. H. Totsuji, T. Kishimoto, Y. Inoue, C. Totsuji, S. Nara, Phys. Lett. A **221**, 215 (1996)
12. H. Totsuji, T. Kishimoto, C. Totsuji, Phys. Rev. Lett. **78**, 3113 (1997)
13. N. Desbiens, P. Arnault, J. Clérouin, Phys. Plasmas **23**, 092120 (2016). https://doi.org/10.1063/1.4963388
14. J.M. Kosterlitz, D.J. Thouless, J. Phys. C Solid St. Phys. **6**, 1181 (1973)
15. K. Strandburg, Rev. Mod. Phys. **60**, 161 (1988)
16. R.A. Quinn, C. Cui, J. Goree, J.B. Pieper, H. Thomas, G.E. Morfill, Phys. Rev. E **53**, 2049 (1996)
17. A. Melzer, A. Homann, A. Piel, Phys. Rev. E **53**, 2757 (1996)
18. V. Nosenko, S.K. Zhdanov, A.V. Ivlev, C.A. Knapek, G.E. Morfill, Phys. Rev. Lett. **103**, 015001 (2009)
19. W.Y. Woon, L. I, Phys. Rev. Lett. **92**, 065003 (2004)
20. B. Klumov, P. Huber, S. Vladimirov, H. Thomas, A. Ivlev, G. Morfill, V. Fortov, A. Lipaev, V. Molotkov, Plasma Phys. Control. Fusion **51**, 124028 (2009)
21. S.A. Khrapak, B.A. Klumov, P. Huber, V.I. Molotkov, A.M. Lipaev, V.N. Naumkin, A.V. Ivlev, H.M. Thomas, M. Schwabe, G.E. Morfill, O.F. Petrov, V.E. Fortov, Y. Malentschenko, S. Volkov, Phys. Rev. E **85**, 066407 (2012)
22. V.N. Naumkin, D.I. Zhukhovitskii, V.I. Molotkov, A.M. Lipaev, V.E. Fortov, H.M. Thomas, P. Huber, G.E. Morfill, Phys. Rev. E **94**, 033204 (2016). http://link.aps.org/doi/10.1103/PhysRevE.94.033204
23. B. Steinmüller, C. Dietz, M. Kretschmer, M.H. Thoma, Phys. Rev. E **97**, 053202 (2018). https://link.aps.org/doi/10.1103/PhysRevE.97.053202
24. J.H. Chu, I. Lin, Phys. Rev. Lett. **72**, 4009 (1994)
25. Y. Hayashi, K. Tachibana, Jpn. J. Appl. Phys. **33**, L804 (1994)
26. H. Thomas, G.E. Morfill, V. Demmel, J. Goree, B. Feuerbacher, D. Möhlmann, Phys. Rev. Lett. **73**, 652 (1994)
27. G.H.P.M. Swinkels, H. Kersten, H. Deutsch, G.M.W. Kroesen, J. Appl. Phys. **88**, 1747 (2000). https://doi.org/10.1063/1.1302993
28. H.R. Maurer, H. Kersten, J. Phys. D Appl. Phys. **44**, 174029 (2011). http://stacks.iop.org/0022-3727/44/i=17/a=174029
29. C. Killer, M. Mulsow, A. Melzer, Plasma Sources Sci. Technol. **24**, 025029 (2015). http://stacks.iop.org/0963-0252/24/i=2/a=025029
30. B. Liu, J. Goree, Y. Feng, Phys. Rev. E **78**, 046403 (2008). https://link.aps.org/doi/10.1103/PhysRevE.78.046403
31. H. Thomas, G.E. Morfill, Nature **379**, 806 (1996)
32. A. Ivlev, U. Konopka, G. Morfill, G. Joyce, Phys. Rev. E **68**, 026405 (2003)
33. A. Ivlev, G. Morfill, Phys. Rev. E **63**, 016409 (2000)
34. S.K. Zhdanov, A.V. Ivlev, G.E. Morfill, Phys. Plasmas **16**, 083706 (2009)
35. B. Liu, J. Goree, Y. Feng, Phys. Rev. Lett. **105**, 085004 (2010)
36. L. Couëdel, V. Nosenko, A.V. Ivlev, S.K. Zhdanov, H.M. Thomas, G.E. Morfill, Phys. Rev. Lett. **104**, 195001 (2010)
37. A. Melzer, Phys. Rev. E **90**, 053103 (2014)
38. V.A. Schweigert, I.V. Schweigert, A. Melzer, A. Homann, A. Piel, Phys. Rev. Lett. **80**, 5345 (1998)
39. F. Melandsø, Phys. Rev. E **55**, 7495 (1997)
40. I.V. Schweigert, V.A. Schweigert, A. Melzer, A. Piel, Phys. Rev. E **62**, 1238 (2000)

Chapter 6
Waves in Weakly Coupled Dusty Plasmas

A paradigm for dynamical processes in many particle system is wave propagation. For dusty plasmas, a large number of wave modes are discussed in literature. Here, we will restrict ourselves to a few fundamental wave types that have been observed in experiments. For a more detailed overview on waves the reader is referred to dedicated monographs [1, 2].

In general, two categories of waves can be identified, namely those which require ordered arrangements of the dust particles (and, hence, strong coupling) and those which do not. In the first category, we find the so-called dust lattice wave (DLW) with its different "polarizations" which will be discussed in the next chapter.

Here, we will start with the discussion of wave types that do not rely on ordered dust. These are, in essence, the dust-acoustic (DAW) and dust ion-acoustic wave (DIAW). The DAW is a wave where the dust particles are the moving species, the DIAW considers the dust as immobile, but the dust influences the wave propagation.

6.1 Dust-Acoustic Wave Dispersion

The dust-acoustic wave has been suggested as one of the first novel wave types in dusty plasmas [3]. The DAW is an analog of the ion-acoustic wave in electron-ion plasmas. The traditional ion-acoustic wave is an electrostatic ion wave where the ions provide the inertia and the electrons provide the shielding. Hence, the ion-acoustic wave oscillates at about the ion plasma frequency

$$\omega_{\mathrm{pi}} = \sqrt{\frac{e^2 n_{\mathrm{i}}}{\varepsilon_0 m_{\mathrm{i}}}}. \tag{6.1}$$

© Springer Nature Switzerland AG 2019
A. Melzer, *Physics of Dusty Plasmas*, Lecture Notes in Physics 962,
https://doi.org/10.1007/978-3-030-20260-6_6

In the DAW, the dust is the mobile species. Hence, now the inertia is given by the dust and both electrons and ions provide the shielding. Hence, the dust-acoustic wave is a very low-frequent wave with wave frequencies of the order of the dust plasma frequency ω_{pd} which, due to the high dust mass, is much less than the ion plasma and electron plasma frequency $(\omega_{pi}, \omega_{pe})$. Analogously to the electron and ion plasma frequency the dust plasma frequency is defined as

$$\omega_{pd} = \sqrt{\frac{Z_d^2 e^2 n_{d0}}{\varepsilon_0 m_d}} \ll \omega_{pi}, \omega_{pe} \,, \tag{6.2}$$

where n_{d0} is the equilibrium (undisturbed) dust density.

Before the mathematical details of the wave motion are described, the wave mechanism is explained (see Fig. 6.1). Consider a wave-like disturbance of the dust density. The dust density disturbance is associated with an enhanced negative space charge due to the negative dust charge. This dust space charge $-Z_d n_d$ will then be immediately shielded by the ambient plasma ions and (to a lesser extent) electrons. In contrast to the ion-acoustic wave, here the shielding is by both electrons and ions. However, the shielding is not complete, the dust charge density fluctuations $-Z_d n_d$ are larger by a very small amount compared to that of electrons and ions $n_i - n_e$. However, this is enough to form negative and positive space charge regions which lead to electric fields E that further drive the wave. The electric field is maximum near the zero-crossing of the dust density perturbations. Therefore, the electric field force pushes the fluctuations in the direction of the electric field and thus the wave propagates.

Fig. 6.1 Comparison of the wave-like dust density disturbance with the ion and electron density fluctuations. It is seen that the dust density fluctuation $Z_d n_d$ is slightly larger than that of electrons and ions $n_i - n_e$

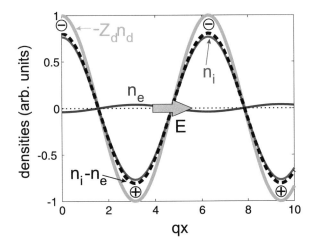

For the derivation of the DAW the equation of continuity (6.3), the momentum equation and Poisson's equation for the dust species are used which can be written as

$$\frac{\partial n_d}{\partial t} + \frac{\partial}{\partial x}(n_d v_d) = 0 \tag{6.3}$$

$$\frac{\partial v_d}{\partial t} + v_d \frac{\partial v_d}{\partial x} + \gamma_d \frac{k_B T_d}{m_d n_d} \frac{\partial n_d}{\partial x} = \frac{Z_d e}{m_d} \frac{\partial \phi}{\partial x} - \beta v_d \tag{6.4}$$

$$\frac{\partial^2 \phi}{\partial x^2} = -\frac{e}{\varepsilon_0}(n_i - n_e - Z_d n_d) . \tag{6.5}$$

Here, we have assumed a simple linear motion along the x-axis. These equations contain few small differences to the usual ion-acoustic wave: the momentum equation (6.4) includes friction with the neutral gas (βv_d), and Poisson's equation (6.5) includes all three charged species, electrons, ions and dust. To solve these equations, the standard approach known from plasma physics textbooks is applied [4]: the dust density and velocity as well as the electron and ion densities are considered as fluctuating quantities with

$$n_d = n_{d0} + \tilde{n}_d$$
$$v_d = 0 + \tilde{v}_d$$
$$\phi = 0 + \tilde{\phi}.$$

The fluctuations are considered to be wave-like, i.e. proportional to $\exp(iqx - i\omega t)$, where q is the wave vector and ω the wave frequency. The electrons and ions are assumed to follow a Boltzmann distribution, namely

$$n_e = n_{e0} \exp\left(\frac{e\tilde{\phi}}{k_B T_e}\right) \simeq n_{e0}\left(1 + \frac{e\tilde{\phi}}{k_B T_e}\right) = n_{e0} + \tilde{n}_e$$

$$n_i = n_{i0} \exp\left(-\frac{e\tilde{\phi}}{k_B T_i}\right) \simeq n_{i0}\left(1 - \frac{e\tilde{\phi}}{k_B T_i}\right) = n_{i0} + \tilde{n}_i$$

Here, n_{e0} and n_{i0} denote the equilibrium (undisturbed) values of the electron and ion density. For the undisturbed densities the quasineutrality condition is fulfilled, i.e.

$$n_{i0} = n_{e0} + Z_d n_{d0},$$

where the dust is assumed to be negatively charged and adds to the electron charge density.

The above three equations can then be written as

$$-i\omega\tilde{n}_d + iq n_{d0}\tilde{v}_d = 0 \tag{6.6}$$

$$-i\omega\tilde{v}_d + iq\gamma_d\frac{k_BT_d}{m_dn_{d0}}\tilde{n}_d = \frac{Z_de}{m_d}iq\tilde{\phi} - \beta\tilde{v}_d \tag{6.7}$$

$$-q^2\tilde{\phi} = -\frac{e}{\varepsilon_0}\left(\tilde{n}_i - \tilde{n}_e - Z_d\tilde{n}_d\right). \tag{6.8}$$

The equilibrium charge densities vanish in Poisson's equation due to quasineutrality. Further, as usual, we have used that for the wave-like fluctuations the spatial and temporal derivations can be replaced by the products with the frequencies and wave vectors ($\partial/\partial t \to -i\omega$; $\partial/\partial x \to iq$). The last equation then becomes

$$-q^2\tilde{\phi} = -\frac{e}{\varepsilon_0}\left(n_{i0}\frac{e\tilde{\phi}}{k_BT_i} + n_{e0}\frac{e\tilde{\phi}}{k_BT_e} - Z_d\tilde{n}_d\right).$$

Applying the quasineutrality again to replace $n_{e0} = n_{i0} - Z_dn_{d0}$ and using the relative dust density $\varepsilon = n_{d0}/n_{i0}$ we get

$$-q^2\tilde{\phi} = -\frac{e^2n_{i0}}{\varepsilon_0 k_BT_i}\left(1 + \frac{T_i}{T_e}[1 - Z_d\varepsilon]\right)\tilde{\phi} + \frac{eZ_d}{\varepsilon_0}\frac{q}{\omega}n_{d0}\tilde{v}_d,$$

where we have used the first equation to replace $\tilde{n}_d$ with $\tilde{v}_d$. From that we can write

$$\tilde{\phi} = -\frac{eZ_d}{\varepsilon_0}\frac{q}{\omega}n_{d0}\tilde{v}_d\frac{\lambda_{D,i}^2}{q^2\lambda_{D,i}^2 + 1 + \frac{T_i}{T_e}[1 - \varepsilon Z_d]}.$$

Inserting this expression for $\tilde{\phi}$ and the first equation into the equation of motion we yield after a few rearrangements

$$\omega^2 + i\beta\omega = \left(\gamma_d\frac{k_BT_d}{m_d} + \varepsilon Z_d^2\frac{k_BT_i}{m_d}\frac{1}{\left(1 + \frac{T_i}{T_e}(1 - \varepsilon Z_d) + q^2\lambda_{D,i}^2\right)}\right)q^2. \tag{6.9}$$

This is the full dispersion relation of the DAW. It contains a number of effects. The first term in the brackets is the dust thermal velocity and the second contains the influence of the electron and ion drive on the dust inertia. This is more clearly seen under the typical assumption of cold dust ($T_d = 0$) and cold ions ($T_i \ll T_e$). Then, the dispersion relation simplifies to

$$\omega^2 + i\beta\omega = \frac{\omega_{pd}^2q^2\lambda_{D,i}^2}{1 + q^2\lambda_{D,i}^2} = \frac{\omega_{pd}^2q^2b^2}{\kappa^2 + q^2b^2}, \tag{6.10}$$

which is the same as for the ion-acoustic wave where the ion properties are replaced by those of the dust and the electron properties by those of the ions. The second expression has been obtained by introducing the screening strength $\kappa = b/\lambda_{D,i}$. Thus the wave frequencies (and wave speeds) decrease with increasing κ.

6.2   Discussion of the Dispersion Relation

The dispersion relation of the DAW (6.10) is shown in Fig. 6.2a. For large wave numbers $q^2\lambda_{D,i}^2 \gg 1$ the wave is not propagating and oscillates at the dust plasma frequency ω_{pd}. For small wave numbers $q^2\lambda_{D,i}^2 \ll 1$ the wave is acoustic $\omega = qC_{DAW}$ with the dust-acoustic sound speed

$$C_{DAW} = \sqrt{\frac{k_B T_i}{m_d} \varepsilon Z_d^2} . \tag{6.11}$$

As for the traditional ion-acoustic wave, the wave speed of the DAW is determined by the temperature of the lighter species (here, the ions, at T_i) and the mass of the heavier (here, the dust, with m_d). The dust-acoustic wave speed also includes the contribution of the dust charge Z_d and the relative dust concentration ε. It is also interesting to note that the governing shielding length is the *ion* Debye length $\lambda_{D,i}$ as the ions are the oppositely charged fluid that predominantly shields the repulsion between the dust particles, see also the discussion in Sect. 6.5.

We now like to analyze the influence of friction on the wave motion. Before doing so we like to remind how instabilities and waves are treated in "usual" plasma physics. A wave that oscillates according to $\exp(iqx - i\omega t)$ becomes unstable when the imaginary part ω_i of the wave frequency $\omega = \omega_r + i\omega_i$ becomes larger than zero.

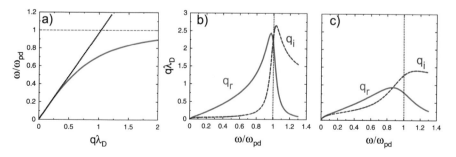

Fig. 6.2 (a) Dispersion relation of the dust-acoustic wave without damping. The solid line is the full dispersion relation, the dotted line indicates the acoustic limit with the dust-acoustic velocity. (b) Dispersion relation with small friction ($\beta = 0.1\,\omega_{pd}$) and (c) with large friction ($\beta = 0.5\,\omega_{pd}$). Here, the solid line refers to the real part of the wave vector and the dashed line to the imaginary part. Note, that in (b) and (c) the axes have been exchanged compared to (a)

Then the wave contains a factor that is increasing exponentially in time. The wave vector q is treated as a real value.

In contrast, when waves are externally driven in a frictional medium the wave frequency ω has to be taken as a real value and, consequently, the wave vector has to be treated as complex $q = q_r + iq_i$, where the real part $q_r = 2\pi/\lambda$ is related to the wave length λ and the imaginary part $q_i = 1/L$ to the damping length L in the system. The damping length L is the distance at which the wave amplitude is reduced to $1/e$. In this situation, we have to determine the real and the imaginary part of the wave vector for each value of the wave frequency. Thus, it is more convenient to plot the dispersion relation as $q_r(\omega)$ and $q_i(\omega)$ instead of the "usual" dispersion $\omega_{r,i}(q)$.

Figure 6.2b,c shows the DAW dispersion for small and large values of the friction coefficient β. For small friction the real part of the wave vector behaves similar to the case of no damping. Close to $\omega = \omega_{pd}$ the wave vector turns over and decreases dramatically towards zero again. In this range the imaginary part of the wave vector jumps from small values, i.e. low damping, to large values. For $\omega > \omega_{pd}$ an overcritically damped DAW is found.

For larger friction constants (Fig. 6.2c) the wave speed ω/q increases and the maximum observable wave number decreases drastically. Moreover, the real and imaginary part of the wave vector are comparable over the entire range: the DAW is found to be strongly damped throughout.

6.3 Experiments on Dust-Acoustic Waves

Dust-acoustic waves have been observed experimentally in weakly [5, 6] and strongly coupled dusty plasma systems [7]. In the weakly coupled system [5, 6], a dc discharge is driven between an anode disk and the chamber walls. The dust particles (aluminum silicate particles with a mean size around 5 μm) are accumulated from a dust tray placed below the anode region. The dust is found to form waves with regions of high and low dust density. The wavelength was about 6 mm at a wave frequency of about 15 Hz (see Fig. 6.3a). In a follow-up experiment with dust of about 1 μm diameter the dispersion relation has been measured by applying a sinusoidal voltage on the anode that drives the wave, see Fig. 6.3b. The corresponding wavelength in the dust system is measured. From the dependence of the wavelength $\lambda = 2\pi/q$ on the driving frequency $f = \omega/(2\pi)$ the dispersion $\omega(q)$ or $q(\omega)$ is derived. The wave is found to show a linear, acoustic dispersion in agreement with the DAW at long wavelengths.

In a different experiment [7], dust-acoustic waves have been driven in a plasma crystal by a sinusoidal voltage on a wire close to the crystal, see Fig. 6.4. The plasma crystal was formed by trapping particles of 9.4 μm diameter in the sheath of an rf discharge in krypton at a gas pressure of 30 Pa. The particle wave motion was observed spatially and temporally resolved by video cameras. For each applied excitation frequency, the corresponding wave length and damping length are

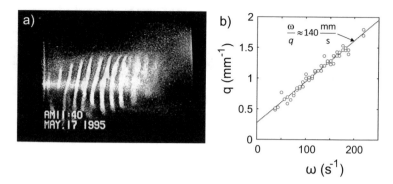

Fig. 6.3 Observation of the DAW in a dc discharge. (**a**) The DAW is seen as regions of high and low dust density in scattered light. Reprinted from [5], with the permission of AIP Publishing. (**b**) Measured dispersion relation $q(\omega)$ of the DAW. Adapted from [6]

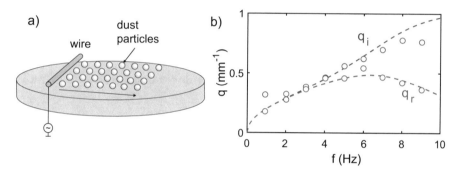

Fig. 6.4 (**a**) Wire excitation of the DAW. The wave is driven by a sinusoidal voltage on the wire close to a plasma crystal. (**b**) Measured dispersion of the DAW: real and imaginary part of the wave vector as a function of wave frequency. The dashed lines correspond to the real and imaginary part of q derived from Eq. (6.10) with $\omega_{pd} = 2\pi \times 9$ Hz, $\lambda_{D,i} = 1.19$ mm and $\beta = 9$ s^{-1}. Reprinted figure with permission from [7]. © (1996) by the American Physical Society

derived. From that the DAW wave dispersion $q_r(\omega)$ and $q_i(\omega)$ has been determined. The measured dispersion relation was found to be in close agreement with a damped DAW, although the system is strongly coupled. Compare the measured dispersion relation in Fig. 6.4b with that of the calculated in Fig. 6.2c.

6.4 Ion-Flow Driven Dust-Acoustic Waves

Dust-acoustic waves do not need to be externally excited. In contrast, self-excited dust-acoustic waves are naturally found in a number of experiments under different conditions (dc or rf discharges in the laboratory or under microgravity) (e.g. [5, 8–12] to list a few), see also Fig. 6.5 and also Fig. 6.3a. In these experiments, the dust

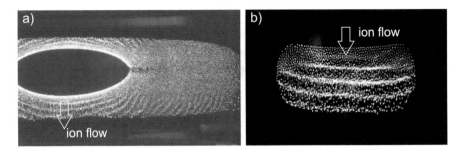

Fig. 6.5 Snap shots of naturally excited DAWs. (**a**) DAW in an rf discharge under microgravity. From [11]. (**b**) DAW in a laboratory rf discharge

is typically confined by an ambipolar or sheath electric field. This electric field also drives an ion flow (actually, essentially any dust cloud in an experiment is subject to ion streaming motion). As the two-stream instability in "ordinary" plasmas excites slow space charge waves, here in a dusty plasma, the ion current excites the slow dust-acoustic waves. For these driven dust-acoustic waves sometimes also the (more adequate) term "dust-density wave" is used, see [13].

The dispersion relation of such an ion-flow driven dust-acoustic wave can be written in simplified form as

$$\varepsilon(\omega, q) = 1 + \frac{1}{q^2 \lambda_{\mathrm{D,e}}^2} - \frac{\omega_{\mathrm{pi}}^2}{\Omega_{\mathrm{i}}(\Omega_{\mathrm{i}} + i\nu_{\mathrm{in}}) - q^2 v_{\mathrm{th,i}}^2} - \frac{\omega_{\mathrm{pd}}^2}{\omega(\omega + i\beta)} = 0, \qquad (6.12)$$

where u_{i} is the ion flow velocity and

$$\Omega_{\mathrm{i}} = \omega - q u_{\mathrm{i}}$$

is the Doppler-shifted frequency that the drifting ion "see". The plasma dielectric function $\varepsilon(\omega, q)$ includes the electron, ion and dust susceptibilities. For the electrons, again, we have taken the low-frequency limit since we are interested in the very low-frequency dust-acoustic waves [compare Eq. (4.8)]. The ion term includes the ion drift motion as well as ion-neutral collisions in form of the collision frequency ν_{in}. For the dust, temperature is neglected (i.e. $v_{\mathrm{th,d}} = 0$), but friction with the neutral gas is included by the Epstein friction coefficient β.

The dispersion relation has, in general, four complex roots for the frequency $\omega = \omega_{\mathrm{r}} + i\omega_{\mathrm{i}}$, but we are interested only in the very low frequency limit $\omega \approx \omega_{\mathrm{pd}} \ll \omega_{pi}, \omega_{pe}$. Figure 6.6 shows the calculated real and imaginary parts of the frequencies for the ion-flow driven waves according to Eq. (6.12) in comparison to the undriven case according to Eq. (6.10).[1]

---

[1]Since we are dealing with self-excited waves, here, we analyze the situation with complex wave frequencies instead of, as above, with complex wave vectors.

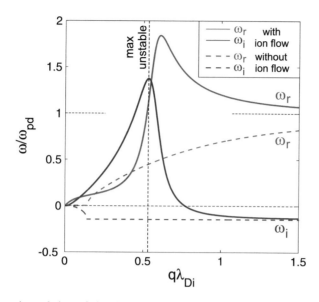

Fig. 6.6 Dispersion relation of the dust-acoustic wave excited by an ion flow (solid lines) according to Eq. (6.12). The wave is most unstable for $q\lambda_{D,i} \approx 0.5$ as indicated by the vertical line. In comparison the dispersion of the undriven DAW according to Eq. (6.10) is also shown (dashed lines). The red and blue lines indicate the real and imaginary part of the wave frequency for a real wave vector q, respectively. The friction coefficient is about $\beta = 0.28\omega_{pd}$ in both cases

First, it is seen that the dispersion for the two cases look quite different. The attainable real wave frequencies are higher for the driven case. They can even somewhat exceed ω_{pd}. The imaginary part of the frequency for the driven wave is positive for a large range of wave vectors q indicating that the wave is (exponentially) growing in time. Hence, the driven waves are unstable despite the presence of friction of the dust particles with the neutral gas (as accounted for by β). Only for very short waves ($q\lambda_{D,i} > 1$) wave damping occurs. The most unstable wave is found for $q\lambda_{D,i} \approx 0.5$ with $\omega_r \approx \omega_{pd}$ under the chosen conditions (which are similar to that of Ref. [9]).

In contrast a wave without the ion-flow drive under the same conditions shows a slow increase of the wave frequency (real part) from zero towards the dust plasma frequency ω_{pd}. The imaginary part is almost constantly negative at $-\beta/2$ indicating the wave damping by neutral gas friction. Very long waves ($q \to 0$) are completely damped.

Hence, the (almost) omnipresent ion flow can excite dust-acoustic waves against the neutral gas damping. The wave dispersion is accordingly modified and quantitatively differs from the simple undriven case, see also [13]. It has also been shown that the dust waves can be driven under an angle relative to the ion motion [11]. The dispersion of these self-excited waves has been used as a diagnostic tool for determining the charge in dense dust particle clouds [14].

6.5 Dust Ion-Acoustic Wave

Next, the dust ion-acoustic wave will be discussed. The dust ion-acoustic wave is the usual ion-acoustic wave with the addition that also an influence of the dust is taken into account. As for the traditional ion-acoustic wave the mobile species in the DIAW is the ions. Hence, the wave frequencies are of the order of the ion plasma frequency $\omega_{pi} \gg \omega_{pd}$ and the dust particles can considered as immobile in the DIAW at these frequencies. The influence of the dust lies only in the reduction of the free electron density since a certain fraction of the electrons is attached to the dust. We apply the equation of continuity, the momentum equation and Poisson's equation for the ions

$$\frac{\partial n_i}{\partial t} + \frac{\partial}{\partial x}(n_i v_i) = 0 \tag{6.13}$$

$$\frac{\partial v_i}{\partial t} + v_i \frac{\partial v_i}{\partial x} = -\frac{e}{m_i}\frac{\partial \phi}{\partial x} \tag{6.14}$$

$$\frac{\partial^2 \phi}{\partial x^2} = -\frac{e}{\varepsilon_0}(n_i - n_e) . \tag{6.15}$$

These equations are exactly those which are used to derive the dispersion relation of the ion-acoustic wave in standard textbooks. Here, in the momentum equation the kinetic pressure of the ions and the friction force are not considered. Poisson's equation does not include the dust as a fluctuating quantity since the dust is immobile. The electrons are treated as Boltzmann distributed

$$n_e = n_{e0} \exp\left(\frac{e\tilde{\phi}}{k_B T_e}\right) \simeq n_{e0}\left(1 + \frac{e\tilde{\phi}}{k_B T_e}\right) = n_{e0} + \tilde{n}_e$$

The only place where the dust properties enter is the quasineutrality condition

$$n_{i0} = n_{e0} + Z_d n_{d0}.$$

Thus, here, the undisturbed electron and ion densities are different since a fraction of the electrons is bound on the dust. The equations are solved analogously as for the DAW. The dispersion relation of the DIAW is then given as

$$\omega^2 = \frac{\omega_{pi}^2 \lambda_{D,e}^2 q^2}{1 + q^2 \lambda_{D,e}^2} = \left(\frac{n_{i0}}{n_{e0}}\right)\frac{k_B T_e}{m_i}\frac{q^2}{1 + q^2 \lambda_{D,e}^2} \tag{6.16}$$

which is that of the pure ion-acoustic wave with the additional factor of $n_{i0}/n_{e0} > 1$. The dispersion relation of the DIAW is shown in Fig. 6.7 in comparison to the usual ion-acoustic wave. The sound speed of the DIAW (ω/q for $q \to 0$) is larger than

Fig. 6.7 Dispersion relation of the DIAW in comparison to the usual ion-acoustic wave (IAW)

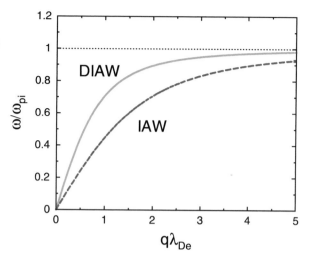

that of the ordinary ion-acoustic wave by the additional factor $(n_{i0}/n_{e0})^{1/2}$, namely

$$C_{\mathrm{DIAW}} = \sqrt{\frac{n_{i0}\,k_B T_e}{n_{e0}\,m_i}} \,. \tag{6.17}$$

Thus, with increasing dust charge density and thus reduced electron density the speed of the DIAW will increase in comparison to the pure ion-acoustic wave. The DIAW has the same maximum frequency, the ion plasma frequency ω_{pi}.

The DIAW can be obtained from the ion-acoustic wave by rescaling the $q\lambda_{D,e}$-axis since for the DIAW the electron Debye length differs from that of the IAW due to the reduced electron density. The traditional interpretation of the ion-acoustic wave speed as that of a sound wave with electron pressure as driver and ion mass density as inertia becomes questionable in the light of Eq. (6.17), see also [13]. In the usual interpretation a reduced electron density would result in less electron pressure and, hence, reduced phase speed in contrast to Eq. (6.17). One also readily sees that in Eq. (6.17) the numerator is not the electron pressure and the denominator is not the ion mass density which is obscured in a quasineutral electron-ion plasma where $n_{e0} = n_{i0}$. As a consequence, a more reasonable interpretation is that the electrons provide the shielding of the mutual ion interaction. Reduced electron density then means reduced shielding and, consequently, stronger ion-ion interaction and, hence, increased phase speed of the wave. We used that picture already in the discussion of the dust-acoustic wave mechanism in Sect. 6.1.

Experimentally, the DIAW has been observed in a Q-machine plasma, where aluminum silicate dust in the micrometer range is constantly fed to the plasma by a rotating drum [15, 16]. It was found that the wave speed of the (dust) ion-acoustic wave increases as expected with the increased dust charge density, see Fig. 6.8. Here, η is a measure of the dust charge density where $\eta = 1$ means no dust and $\eta \to 0$

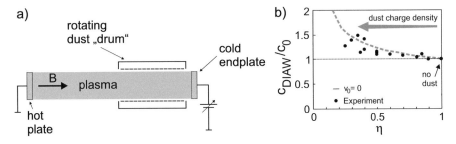

Fig. 6.8 (**a**) Experimental setup for the observation of the DIAW. The dust is immersed into the plasma by a rotating dust "drum". (**b**) Measured phase velocity of the DIAW normalized to the phase velocity without dust c_0. The dust charge density increases from right to left where $\eta = 1$ means no dust. The line is the expected increase of the DIAW sound speed for zero ion drift velocity. Reprinted and adapted from [15]. © (1996), with permission from Elsevier

means increasing dust charge density. This increase of the wave speed has been taken as an indication of the existence of the DIAW.

6.6 Other Wave Types

In literature, additional waves types are discussed for weakly coupled dusty plasmas under a magnetic field, see also Chap. 9. Among those are the dust-cyclotron wave and the dust ion-cyclotron. The dust cyclotron wave and dust ion-cyclotron wave are, similar to the situation discussed above, again the exact twins of the "usual" ion-cyclotron waves. This ion-cyclotron wave dispersion is known to be

$$\omega^2 = \omega_{ci}^2 + \frac{k_B T_e}{m_i} q^2$$

with the ion cyclotron frequency $\omega_{ci} = eB/m_i$ at the magnetic field strength B.

In the dust-cyclotron wave the dust is the mobile species. The ions in the ion-cyclotron wave are replaced by the dust and both electrons and ions take the role from the electrons. Consequently, the wave dispersion relation of the dust-cyclotron wave readily reads as

$$\omega^2 + i\beta\omega = \omega_{cd}^2 + \varepsilon Z_d^2 \frac{k_B T_i}{m_d} q^2$$

where $\omega_{cd} = Z_d eB/m_d$ is the cyclotron frequency of the dust. It is the same dispersion relation as for the DAW, only shifted by the cyclotron frequency. So, the ion-acoustic wave speed of the ion-cyclotron wave is replaced by the dust-acoustic wave speed for the dust-cyclotron wave.

Similarly, in the dust ion-cyclotron wave the dust is immobile and influences the dispersion relation only through the quasineutrality condition. Thus, the dust ion-cyclotron dispersion includes the term $n_{i0}/n_{e0} > 1$ to account for the electrons bound to the dust. The dispersion then is

$$\omega^2 = \omega_{ci}^2 + \left(\frac{n_{i0}}{n_{e0}}\right)\frac{k_B T_e}{m_i}q^2.$$

As for the dust ion-acoustic wave, dust ion-cyclotron waves have been driven in a magnetized Q-machine plasma [16]. There, an increased amplitude of the ion-cyclotron wave has been observed with increased dust charge density. This is not a direct measurement of the increased phase speed as in the case of the DIAW, but it demonstrates that the presence of dust allows an easier excitation of the ion-cyclotron wave.

Moreover, a vast number of other dust-influenced waves has been discussed in literature, see also [1, 2]. However, experimental confirmations of such waves are rare. In addition, nonlinear properties of the various wave types have also been analyzed analytically and numerically. Again, only a few experiments are available. We thus limit the presentation of dust waves in weakly coupled systems to the examples mentioned here.

References

1. F. Verheest, *Waves in Dusty Space Plasmas* (Kluwer Academic Publishers, Dordrecht, 2000)
2. P.K. Shukla, A.A. Mamun, *Introduction to Dusty Plasma Physics* (Institute of Physics Publishing, Bristol, 2002)
3. N.N. Rao, P.K. Shukla, M.Y. Yu, Planet. Space Sci. **38**, 543 (1990)
4. A. Piel, *Plasma Physics: An Introduction to Laboratory, Space, and Fusion Plasmas* (Springer, Heidelberg, 2010)
5. A. Barkan, R.L. Merlino, N. D'Angelo, Phys. Plasmas **2**, 3563 (1995)
6. C. Thompson, A. Barkan, N. D'Angelo, R.L. Merlino, Phys. Plasmas **4**, 2331 (1997)
7. J.B. Pieper, J. Goree, Phys. Rev. Lett. **77**, 3137 (1996)
8. H. Prabhakara, V. Tanna, Phys. Plasmas **3**, 3176 (1996)
9. J.D. Williams, J. Edward Thomas, L. Marcus, Phys. Plasmas **15**, 043704 (2008)
10. S. Ratynskaia, S. Khrapak, A. Zobnin, M.H. Thoma, M. Kretschmer, A. Usachev, V. Yaroshenko, R.A. Quinn, G.E. Morfill, O. Petrov, V. Fortov, Phys. Rev. Lett. **93**, 085001 (2004)
11. A. Piel, M. Klindworth, O. Arp, A. Melzer, M. Wolter, Phys. Rev. Lett. **97**, 205009 (2006)
12. T.M. Flanagan, J. Goree, Phys. Plasmas **17**, 123702 (2010)
13. A. Piel, AIP Conf. Proc. **1397**(1), 50 (2011). https://aip.scitation.org/doi/abs/10.1063/1.3659739
14. B. Tadsen, F. Greiner, S. Groth, A. Piel, Phys. Plasmas **22**(11), 113701 (2015). http://scitation.aip.org/content/aip/journal/pop/22/11/10.1063/1.4934927
15. A. Barkan, N. D'Angelo, R.L. Merlino, Planet. Space Sci. **44**, 239 (1996)
16. R.L. Merlino, A. Barkan, C. Thompson, N. D'Angelo, Phys. Plasmas **5**, 1607 (1998)

Chapter 7
Waves in Strongly Coupled Dusty Plasmas

In this section, dust lattice waves (DLW) will be discussed. As the name suggests, the dust lattice wave requires the particles to be ordered in a crystal lattice. We thus need a strongly coupled dust system.

The dust lattice waves will be treated here mostly for a 2D system trapped in the sheath of a discharge. Then, the dust lattice wave has three different "polarizations", namely compressional, shear and transverse modes (see Fig. 7.1). For the compressional (longitudinal) mode, the particle motion is along the wave propagation direction and leads to compression and rarefaction of the dust component. In the shear mode, the dust motion is perpendicular to the wave propagation, but inside the 2D crystal plane. The transverse mode also describes particle motion perpendicular to the wave propagation, but here the dust motion is an out-of-plane motion and thus requires the consideration of the vertical confinement of the dust. These three wave types have been observed in experiments and will be presented in the following.

7.1 Compressional Mode in 1D

We will start to illustrate the dispersion relation of the 2D-DLW from the simpler model of a 1D chain of dust particles, see Fig. 7.2. In an ordered linear chain the dust particles have equidistant equilibrium positions $X_n = nb$, where b is the interparticle distance. Neighboring dust particles are considered to be connected by springs of spring constant k. These "spring constants" reflect of course the repulsive interaction between the dust particles as described below. The force on the n-th particle is $F = k(x_{n-1} - x_n)$ due to the spring to the left neighbor and $F = k(x_{n+1} - x_n)$ due to the right neighbor. The equation of motion for the n-th particle then is

$$m_d \ddot{x}_n - m_d \beta \dot{x}_n = k(x_{n-1} - 2x_n + x_{n+1}) , \tag{7.1}$$

© Springer Nature Switzerland AG 2019
A. Melzer, *Physics of Dusty Plasmas*, Lecture Notes in Physics 962,
https://doi.org/10.1007/978-3-030-20260-6_7

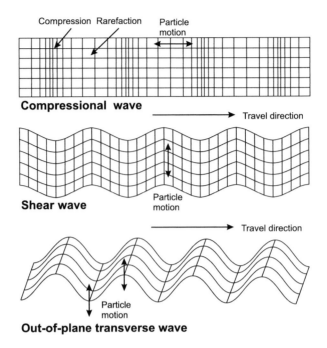

Fig. 7.1 Compressional wave, shear wave and transverse wave

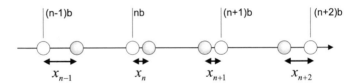

Fig. 7.2 Sketch of a linear dust arrangement with longitudinal particle displacements. The white circles indicate the equilibrium positions in the chain, the bluish the instantaneous positions in the traveling wave

where x_n is the elongation of the n-th particle from its equilibrium position X_n. Here, also friction with the neutral gas is included. Using the ansatz for waves on a linear chain $x_n = A \exp(inqb - i\omega t)$ with wave vector q and wave frequency ω the equation of motion becomes

$$-m_d \omega^2 - i m_d \beta \omega = k(e^{iqb} + e^{-iqb} - 2) = 2k(\cos qb - 1) = -4k \sin^2 \left(\frac{qb}{2} \right)$$

and the dispersion relation (see also Sect. 4.5)

$$\omega^2 + i\beta\omega = 4 \frac{k}{m_d} \sin^2 \left(\frac{qb}{2} \right) \tag{7.2}$$

is obtained. This is the well-known dispersion relation of waves on a linear chain which is also very familiar in condensed matter physics.

Now, the spring constant k has to be related to the repulsive interaction between the dust particles. From mechanics it is known that the spring constant can be seen as the second derivative of the interaction energy. This yields for a Debye-Hückel interaction, Eq. (4.4), at the interparticle distance b

$$k = \left.\frac{d^2 V}{dr^2}\right|_{r=b} = \frac{Z_d^2 e^2}{4\pi \varepsilon_0 b^3} e^{-\kappa} \left(2 + 2\kappa + \kappa^2\right), \tag{7.3}$$

where the screening strength $\kappa = b/\lambda_D$ has been used. The corresponding dispersion can then be written as

$$\omega^2 + i\beta\omega = \frac{Z_d^2 e^2}{\pi \varepsilon_0 m_d b^3} e^{-\kappa} \left(2 + 2\kappa + \kappa^2\right) \sin^2 \left(\frac{qb}{2}\right), \tag{7.4}$$

In a final step, the dispersion relation can be extended to include also the influence of many neighbors. Therefore, simply the "springs" to all other neighbors at distance ℓb have to be considered yielding [1]

$$m_d \ddot{x}_n - m_d \beta \dot{x}_n = \sum_{\ell=1}^{\infty} k(\ell b)(x_{n-\ell} - 2x_n + x_{n+\ell}) .$$

The full 1D dispersion relation then is given by [1]

$$\omega^2 + i\beta\omega = \frac{1}{\pi}\omega_{pd}^2 \sum_{\ell=1}^{\infty} \frac{e^{-\ell\kappa}}{\ell^3} \left(2 + 2\ell\kappa + \ell^2\kappa^2\right) \sin^2 \left(\frac{\ell q b}{2}\right), \tag{7.5}$$

where the dust plasma frequency

$$\omega_{pd}^2 = \frac{Z_d^2 e^2}{\varepsilon_0 m_d b^3} \tag{7.6}$$

has been introduced for the strongly coupled case by identifying $n_d \approx b^{-3}$ in (6.2). It should, however, be noted that the dust plasma frequency is used here only because of the mathematical similarity. It does not correspond to a collective oscillation of the dust particles relative to a stationary background. Rather, here, this frequency is related to the Debye frequency in solid-state physics.

The dispersion relation of the compressional 1D dust lattice wave is shown in Fig. 7.3 for different values of κ. The dispersion relation has approximately the form of a sine function. For small wave numbers (long wavelengths) the wave is acoustic ($\omega \propto q$), whereas the dispersion bends over for larger wave numbers and finally reaches a maximum where $\partial \omega / \partial q \approx 0$. The maximum wave number is given

Fig. 7.3 Dispersion relation
of the 1D dust lattice wave
for different values of the
screening strength κ at zero
damping

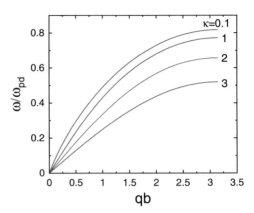

by $qb = \pi$ which corresponds to a wave length $\lambda = 2b$. This wavelength is the
shortest possible wave length on a string of particles and corresponds to 180-degree
out-of-phase motion between nearest neighbor particles. As for the DAW, the wave
frequencies of the DLW decrease with increasing κ.

The sound speed of the 1D compressional DLW is given by

$$c = \lim_{q \to 0} \frac{\partial \omega}{\partial q} = \lim_{q \to 0} \frac{\omega}{q} = \sqrt{\frac{Z_d^2 e^2}{2\pi \varepsilon_0 m_d b}} \sqrt{\frac{1}{2} \sum_{\ell=1}^{\infty} \frac{e^{-\ell \kappa}}{\ell} \left(2 + 2\ell \kappa + \ell^2 \kappa^2\right)} = c_0 f(\kappa),$$

(7.7)

where

$$c_0 = \sqrt{\frac{Z_d^2 e^2}{2\pi \varepsilon_0 m_d b}}$$

(7.8)

is a measure of the sound speed of dust lattice waves in plasma crystals[1] and

$$f(\kappa) = \sqrt{\frac{1}{2} \sum_{\ell=1}^{\infty} \frac{e^{-\ell \kappa}}{\ell} \left(2 + 2\ell \kappa + \ell^2 \kappa^2\right)}$$

(7.9)

is a function of κ only and takes into account the crystal order.

The effect of damping is shown in Fig. 7.4 where the dispersion relation (7.5) is
solved for an excited wave, i.e. with a real ω and a complex $q = q_r + iq_i$ (compare
also Sect. 6.2). Now, here, the resulting real and imaginary part of the wave vector
are shown as a function of wave frequency. An analytical solution exists when only

---

[1] c_0 is the sound speed when (somewhat self-contradictory) only nearest neighbors ($\ell = 1$) in the
limit $\kappa \to 0$ are taken into account.

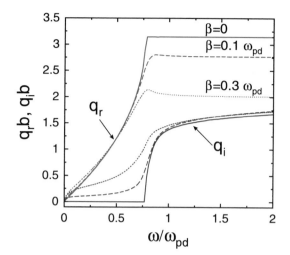

Fig. 7.4 Dispersion relation of the 1D dust lattice wave for different values of the damping strength β at $\kappa = 1$

nearest neighbor interaction ($\ell = 1$) in Eq. (7.5) is taken account. Otherwise the solution has to be found numerically. Figure 7.4 shows the case of many neighbor interaction. It can be seen that with increased damping the maximum real wave vector is reduced. In the same way, the imaginary part (corresponding to the inverse damping length) increases. Similar to the dispersion relation of the DAW, in the DLW the imaginary wave vector grows dramatically when the wave frequency comes close to the maximum wave frequency $\omega_{max} \approx 0.8\omega_{pd}$ (compare Fig. 7.3). Heavily damped waves are found when the wave enters the frequency regime where $\omega > \omega_{max}$.

7.2 Dust Lattice Waves in 2D

The dispersion relation of 1D waves can be extended to the case of a two-dimensional lattice in a straightforward manner for the compressional mode as well as for the shear mode [2, 3].

There, one has to take into account the hexagonal order of the 2D lattice. For the compressional wave, the dispersion relation then reads [2, 3]

$$\omega^2 + i\beta\omega = \frac{1}{2\pi}\omega_{pd}^2 \sum_{\ell,m=1}^{\infty} \frac{e^{-\tilde{r}\kappa}}{\tilde{r}^5} \left(\ell^2\left(3 + 3\tilde{r}\kappa + \tilde{r}^2\kappa^2\right) - \tilde{r}^2(1 + \tilde{r}\kappa)\right)\sin^2\left(\frac{\ell q b}{2}\right),$$

(7.10)

where $\tilde{r} = \sqrt{\ell^2 + m^2}$. The term after the sum is just the second derivative of the Yukawa potential in x-direction at the lattice site (ℓ, m). Here, it is assumed that the wave propagates in x-direction (along $\mathbf{a}_1$ in Fig. 7.5). Thus, the force (related to

Fig. 7.5 Hexagonal lattice with the lattice vectors a_1 and a_2. The expression in brackets (ℓ, m) gives the 2D particle position in units of the interparticle distance b

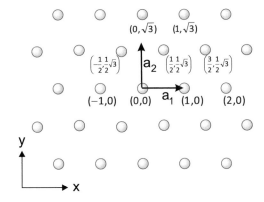

the derivative) and the corresponding particle motion is in the same direction as the wave. Hence, this describes a compressional (longitudinal) wave mode.

For the shear wave one also has to take the second derivative, but in the y-direction (along $\mathbf{a_2}$), when the wave is propagating in the x-direction. The force and the motion are in the y-direction whereas the wave propagates in the x-direction. Hence, the wave is a shear wave. (Strictly speaking, this type of wave can be termed transverse wave. However, we like to reserve the term "transverse wave" for the out-of-plane wave discussed in Sect. 7.7).

That yields [2, 3]

$$\omega^2 + i\beta\omega = \frac{1}{2\pi}\omega_{pd}^2 \sum_{\ell,m=1}^{\infty} \frac{e^{-\tilde{r}\kappa}}{\tilde{r}^5}\left(m^2\left(3 + 3\tilde{r}\kappa + \tilde{r}^2\kappa^2\right) - \tilde{r}^2\left(1 + \tilde{r}\kappa\right)\right)\sin^2\left(\frac{\ell q b}{2}\right).$$
$$(7.11)$$

The shear wave character is seen from the fact that the term after the sum includes m (denoting the lattice site in y-direction) whereas the sin-expression includes ℓ (denoting the lattice site in x-direction along wave propagation).

The computed dispersion relation of the compressional and shear 2D dust lattice wave is shown in Fig. 7.6. The compressional mode has a form that is very similar to the 1D-case: For long wavelengths $qb \ll 1$ the dispersion is acoustic. For shorter wave lengths the compressional mode becomes dispersive and attains a maximum near $qb \approx \pi$. In contrast, the shear mode is nearly acoustic for all wavelengths. Thus, the shear mode is only little dispersive.

It is readily seen here, that the sound speed of the compressional mode is much larger than that of the shear mode. The sound speeds of the compressional and shear mode are given by [2]

$$c_{comp} = c_0 f_{comp}(\kappa) \qquad c_{shear} = c_0 f_{shear}(\kappa) \qquad (7.12)$$

Fig. 7.6 Dispersion relation of the 2D dust lattice wave for different values of the screening strength κ (without damping). The solid lines represent the compressional mode, the dashed lines the shear mode

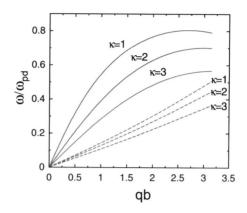

where

$$f_{\text{comp}} = \sqrt{\frac{1}{2}\sum_{\tilde{r}}^{\infty}\frac{e^{-\tilde{r}\kappa}}{\tilde{r}}\left(\frac{15}{8}(1+\tilde{r}\kappa)+\frac{9}{8}\tilde{r}^2\kappa^2\right)}$$

and

$$f_{\text{shear}} = \sqrt{\frac{1}{2}\sum_{\tilde{r}}^{\infty}\frac{e^{-\tilde{r}\kappa}}{\tilde{r}}\left(-\frac{3}{8}(1+\tilde{r}\kappa)+\frac{3}{8}\tilde{r}^2\kappa^2\right)}$$

where the summation is over all possible distances in the hexagonal lattice $\tilde{r}$ and c_0 is given by Eq. (7.8). Comparing the sound speed of the compressional 1D and 2D waves one finds only slight differences, e.g. the factor 15/8 instead of 2 and 9/8 instead of 1. These factors are the effect of the hexagonal lattice structure. The sound speed of the shear mode contains a negative term which substantiates that the sound speed of the shear mode is smaller than that of the compressional. This dispersion holds for finite values of the screening strength κ. For pure Coulomb interaction $\kappa = 0$ the sum in the compressional dispersion relation (7.10) would diverge. For pure Coulomb interaction it is found that $\omega \propto \sqrt{q}$ for long wavelengths (and thus the sound speed $c = \omega/q \to \infty$ for $q \to 0$).

Moreover, the wave dispersion has been discussed for a compressional wave propagating along $\mathbf{a}_1$ and a shear wave along $\mathbf{a}_2$. In the general case there can be any arbitrary angle θ of the wave propagation $\mathbf{q}$ relative to the lattice orientation $\mathbf{a}_1$ which has an influence on the dispersion relation [3]. However, that influence only manifests in the very short wavelength regime $qb > 2.5$ when the exact position of the nearest neighbors is probed by the wave.

7.3   Compressional 1D Dust Lattice Waves: Experiment

Compressional dust lattice waves in 1D and 2D systems have been identified in experiments and the dispersion has been measured by Homann et al. [4, 5]. Here, we start with the presentation of the 1D results.

A linear chain of dust particles has been realized by placing a rectangular barrier on the electrode (see Fig. 7.7a and Sect. 8.1). Then, particles of $9.5\,\mu m$ diameter have been dropped into the barrier and a chain is formed. Waves in the dust chain have been excited by focusing a laser beam onto the first particle. The laser beam is then switched "on" and "off" periodically. The first particle is moved by the laser and a wave is launched into the linear chain of dust particles. The oscillations are clearly seen in the video snap shots (Fig. 7.7b). It is also obvious that the amplitude of the wave decreases with distance in the chain due to friction with the neutral gas. The wave motion is then analyzed directly from tracking the particles in the video data in terms of phase and amplitude. Therefore the oscillations of each particle in the chain relative to its equilibrium position is identified (see Fig. 7.8a). It is immediately seen that the phase of the oscillation progresses from one particle to next. Correspondingly also the oscillation amplitude decreases. The phases and amplitudes of all particles are then derived and analyzed as a function of equilibrium position (Fig. 7.8b, c). The observed linear dependence of the phase with particle position directly reflects the wave length λ and thus the real part of the wave vector $q_r = 2\pi/\lambda$. The amplitude of the wave is found to decrease exponentially with distance from the first particle. This determines the damping length L and

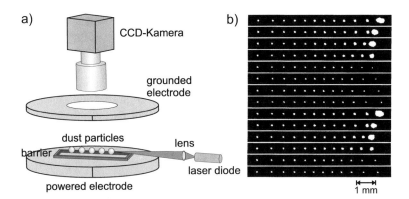

Fig. 7.7 (**a**) Experimental setup for the excitation of 1D dust lattice waves in a chain of dust particles. (**b**) Snap shots of the dust chain oscillation. The first particle shows strong "blooming" when hit by the laser beam. One can see that a wave is propagating into the dust chain. From [4]

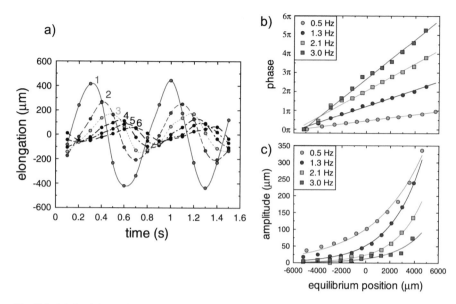

Fig. 7.8 (**a**) Particle oscillation relative to the equilibrium position as a function of time. The numbers denote the particle in the chain, where the first particle is the one hit by the beam. (**b**) Phase and (**c**) amplitude of the particle oscillation derived from (**a**) as a function of position in the chain for different excitation frequencies. From [4]

correspondingly the imaginary part of the wave vector $q_i = 1/L$. Thus, the wave in the chain propagates as

$$x = A \exp(iqx - i\omega t) = A \exp(i(q_r + iq_i)x - i\omega t)$$
$$= A \exp(iq_r x - i\omega t) \exp(-q_i x),$$

where A is the oscillation amplitude of the first particle and x measures the distance from the first particle. The wave vector q_r and q_i depend on the wave frequency ω.

By measuring the imaginary and real part of the wave vector in such a way for different wave frequencies, the entire dispersion relation is identified. Figure 7.9a shows the wave vectors as a function of wave frequency. The measured dispersion relation is then compared with the dispersion relation of the dust-acoustic wave according to Eq. (6.10) and of the dust lattice wave according to Eq. (7.5). It is clearly seen that the DAW dispersion does not match the measured dispersion relation whereas the DLW dispersion is in very good agreement. Thus, the 1D dust chain does not exhibit DAW-like wave motion. Here, the dispersion is clearly of DLW type.

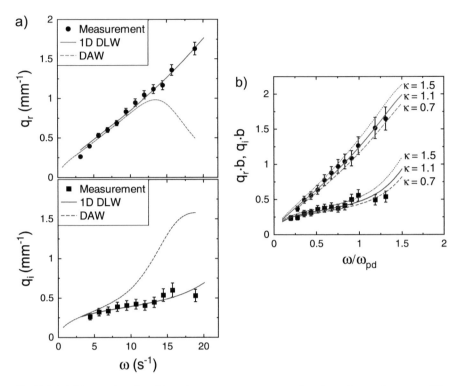

Fig. 7.9 (a) Measured real and imaginary wave vectors as a function of wave frequency. The measured dispersion relation is compared with the dispersion relation of the DAW and DLW. (b) Measured dispersion relation in comparison with the DLW dispersions for various values of κ. In this Figure, normalized units ω/ω_{pd} (using the dust charge $Z_d = 14,000$ measured from the resonance technique) and qb (using the measured interparticle distance $b = 930\,\mu m$) have been used. From [4]

The measured dispersion relation also allows to determine the screening strength κ. This is demonstrated in Fig. 7.9c where the measured dispersion is shown in comparison with theoretical curves for a range of κ-values. Best agreement between experiment and theory is obtained for a screening strength of $\kappa = 1.1$, reasonable agreement is found in the range between $\kappa = 0.7$ and 1.5, thus

$$\kappa = b/\lambda_D = 1.1 \pm 0.4. \tag{7.13}$$

This means that the screening length λ_D in the dust system is of the order of the interparticle distance b (or the other way round). In this case

$$\lambda_D \approx b = 930\,\mu m.$$

This value of the screening length is of the order of the electron Debye length. We will return to the discussion of the screening problem below.

7.4   Compressional 2D Dust Lattice Waves: Experiment

Experiments which are similar to those described above have been performed to measure the dispersion relation of the compressional DLW in two dimensions [5]. In the 2D case, the laser beam of an Argon ion laser was expanded into a line focus and directed onto the first row of particles in a 2D plasma crystal. By periodic modulation of the laser power a plane wave was launched in the plasma crystal (see Fig. 7.10a). Again, the wave motion of the dust was analyzed in terms of the phase and amplitude (Fig. 7.10b, c) as a function of distance from the excitation region. Similarly, the linear phase dependence on equilibrium particle position directly reflects the wavelength λ and the amplitude decrease with position the damping length L for a given excitation frequency. In that way the dispersion relation of the wave has been measured (Fig. 7.10d, e) in terms of the wave vector $q_r = 2\pi/\lambda$ and $q_i = 1/L$. The measured dispersion was found to show good agreement with the expected 2D DLW dispersion. From this comparison the screening strength is determined to be $\kappa = 1\pm0.3$. So, also in 2D the interparticle distance b and shielding length λ_D are comparable and are found to be close to the electron Debye length.

7.5   Shear 2D Dust Lattice Waves: Experiment

Shear dust lattice waves have been observed by Nunomura et al. [6] also using laser beam excitation. There, a short pulse of a narrow laser beam excites a row of particles along the direction of the beam (see Fig. 7.11). The velocity pulse created by the beam propagates in a direction perpendicular to the beam. Hence, the dust particle motion and pulse travel direction are perpendicular to each other, thus a shear wave is observed here. The outward velocity of the pulse ($c \approx 6$ mm/s) is much smaller than for a compressional pulse and in agreement with the acoustic velocity of the shear wave, see Eq. (7.12).

Further, Nunomura et al. [7] have investigated shear and compressional waves in great detail by laser excitation. There, they also have studied the wave propagation along different lattice orientations and found very good agreement with the theoretical DLW dispersion relation. A further refined version of their measurement of the different branches [8, 9] will be described below in Sect. 7.8.

From this detailed measurement of the dispersion relation, Nunomura et al. [7] did not only extract the screening strength, but also the dust charge. The screening strength was found to be close to $\kappa = 1$ in agreement to the previous experiments. The dust charge was extracted from the wave frequencies to be in the range between $Z_d = 11,000$ and $Z_d = 15,000$ for particles of 4.04 μm radius which is very compatible with the values obtained from the resonance technique in Sect. 3.8.1.

This wave method, however, has the great advantage that the wave directly probes the particle-particle interaction (characterized by the dust charge Z_d and the screening strength κ) without requiring a knowledge of the (difficult to measure)

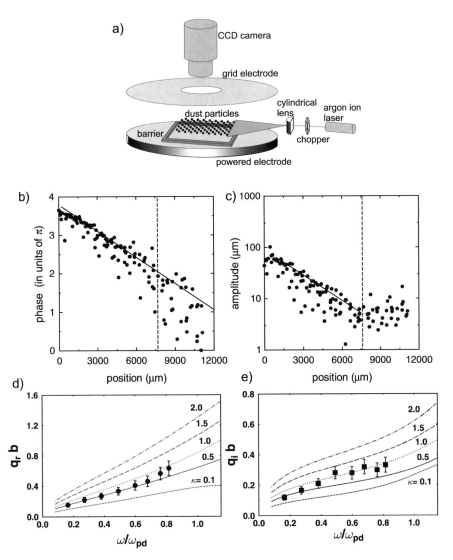

Fig. 7.10 (**a**) Scheme of the experimental setup for the excitation of 2D dust lattice waves. (**b**) Phase and (**c**) amplitude of the dust particle motion as a function of distance from the excitation region for an excitation frequency of 2.8 Hz. (**d**, and **e**) Real and imaginary wave vector as a function of frequency. The symbols denote the experimental data. The lines indicate the dispersion relation of the 2D DLW for various values of the screening strength κ. From [5]

a)   b)

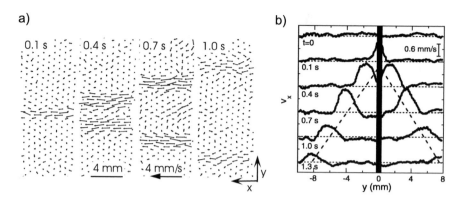

Fig. 7.11 Shear dust lattice waves. (**a**) Dust particle velocity vectors at certain time steps after a laser beam pulse. The initial laser beam pushed the particles in the central region from right to left. (**b**) Velocity profiles perpendicular to the beam direction. The central bar indicates the excitation region. Reprinted figure with permission from [6]. © (2000) by the American Physical Society

plasma parameters as for the resonance technique. Hence, the determination of the wave dispersion allows to reliably determine the relevant dust parameters.

7.6 Mach Cones

When an object moves through a medium with a velocity faster than the sound speed in that medium a V-shaped disturbance, the Mach cone, is excited. This phenomenon is well known, e.g. from the sonic boom behind a plane at supersonic velocity. Similarly, Mach cones can be observed in dusty plasmas using objects faster than the acoustic speed of the DLW.

The mechanism of Mach cone formation is illustrated in Fig. 7.12e. The supersonic object moves at a velocity V that creates disturbances on its way at any instant. The disturbances propagate through the medium with the sound speed c. Since $c < V$ the front of the disturbance lags behind the object, thus forming the well-known V-shaped front, the Mach cone. The Mach cone has an opening angle μ that satisfies the relation

$$\sin \mu = \frac{c}{V} . \tag{7.14}$$

Thus from a single measurement of the opening angle μ the sound speed of the DLW is readily obtained from which other parameters like particle charge or screening strength can be determined.

Mach cones in dusty plasmas have first been observed by Samsonov et al. [10, 11]. There, dust particles which accidentally are trapped below the actual 2D plasma crystal layer are found to move at large, supersonic, speeds at low gas pressure. The

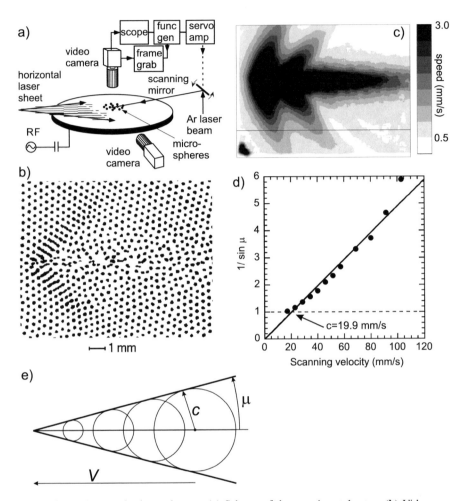

Fig. 7.12 Mach cones in dusty plasmas. (**a**) Scheme of the experimental setup. (**b**) Video snap shot of the dust crystal. The laser spot moves from right to left. A V-shaped disturbance is clearly observable. (**c**) Gray-scale map of the dust particle velocities. (**d**) Test of the Mach cone relation. Plot of $1/\sin \mu$ of the measured cone angle μ as a function of laser spot velocity V. From this, the sound speed is measured to be $c = 19.9$ mm/s. (**e**) Sketch of the Mach cone formation. From [12]

disturbance by these lower fast-moving particles excites a Mach cone in the upper plasma crystal layer.

In a different experiment [12] Mach cones in plasma crystals have been generated using the focal spot of a laser beam that was moved at supersonic speeds V through the crystal using a moving galvanometer scanning mirror (Fig. 7.12a). This laser technique allows the formation of Mach cones in a repetitive and controllable manner.

Figure 7.12b shows a video snap shot where the Mach cone created by the laser spot is easily seen. The opening angle here is about $\mu \approx 45°$. A more detailed picture is obtained when investigating the absolute values of the dust particle velocities. The particle speeds are shown in a gray-scale plot in Fig. 7.12c. Here, a strong first Mach cone is easily seen. However, additional secondary and tertiary Mach cones are also observable. The first strong cone is just the expected behavior for the Mach cone as described above. The additional features arise from the dispersive nature of the DLW at shorter wavelengths [13]. The laser spot creates wave disturbances at all wavelengths. Due to dispersion short wavelength waves travel at a different velocity than those at long wavelengths. This makes the picture described in Fig. 7.12e more complicated: Like the wave pattern of a moving ship (which is not a Mach cone as described above), the secondary and tertiary Mach cones can be interpreted as interference patterns of the waves launched by the moving laser beam.

From the measurement of the opening angle μ of the first Mach cone at various laser spot velocities the Mach cone relation is verified (see Fig. 7.12d). The sound speed measured here is about 20 mm/s which is the expected sound speed of a compressional DLW according to Eq. (7.12). This small value again demonstrates that the dynamic processes in dusty plasmas occur on a long time scale.

The Mach cone in Fig. 7.12 is a compressional Mach cone due to excitation of compressional waves. Shear Mach cones by the excitation of shear waves have been demonstrated by Nosenko et al. [14]. Shear Mach cones are observed at much lower laser spot velocities V due to the much smaller acoustic velocity of the shear waves (about 5 mm/s in that particular experiment, compare also Eq. (7.12)).

Mach cones are discussed to be used as a diagnostic tool: Mach cones might be observable in the rings of Saturn by the Cassini spacecraft [15]. In Saturn's rings, large boulders moving in Keplerian orbits have supersonic speeds relative to the smaller dust particles which move at speeds determined by their electrostatic interactions with Saturn's plasma environment (see Sect. 12.5). The observation of Mach cones would allow detailed studies of the plasma conditions in the rings, but, at present, Mach cones in Saturn's rings have not been reported. Nevertheless, Mach cones have been observed in extended 3D dust clouds under microgravity allowing to probe the dust parameters there [16, 17].

7.7 Transverse (Out-of-Plane) Dust Lattice Waves

So far, we have dealt with waves and particle motions in the plane of the 2D plasma crystal. We now turn to the transverse dust lattice wave. There, a wave-like, out-of-plane (vertical) particle motion is expected where the wave propagation direction is in the crystal plane. Such vertical displacements are to be stabilized against the Coulomb repulsion of the particles by the vertical confinement potential. The external confinement is essential for this type of wave, otherwise the wave would not be stable.

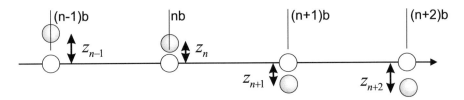

Fig. 7.13 Sketch of a linear dust arrangement with transverse particle displacements

For the dispersion of the transverse DLW vertical displacements z_n in a 1D chain of particles are considered (see Fig. 7.13). The equation of motion then reads

$$m_d\ddot{z}_n - m_d\beta\dot{z}_n + m_d\omega_0^2 z_n = F_z(z_n - z_{n-1}) + F_z(z_n - z_{n+1}) , \qquad (7.15)$$

where ω_0 is the strength of the vertical confinement [compare Eq. (3.22)] and F_z is the vertical interaction force between neighboring particles. Using $\delta z = z_n - z_{n\pm1} \ll b$, the interparticle distance is $r = \sqrt{\delta z^2 + b^2} \approx b$ and the force given by

$$F_z = F(r)\frac{\delta z}{r} \approx F(b)\frac{\delta z}{b} = \frac{Z_d^2 e^2}{4\pi\varepsilon_0 b^3}e^{-\kappa}(1+\kappa)\,\delta z ,$$

where F_z is the vertical component of the Coulomb force

$$F(r) = -\frac{\partial V(r)}{\partial r} = \frac{Z_d^2 e^2}{4\pi\varepsilon_0 b^2}e^{-\kappa}(1+\kappa) .$$

The fact that here the first derivative of the particle interaction energy V is involved (in contrast to the second derivative for the compressional and shear wave) states that the equilibrium situation is unstable in the absence of the vertical confinement ω_0. For the repulsive interaction between the dust particles a small transverse elongation would grow indefinitely. The transverse oscillations are only stabilized by the additional counter-force due to the external confinement.

Following the above procedure, the dispersion relation of the transverse DLW is given by [18]

$$\omega^2 + i\beta\omega = \omega_0^2 - \frac{1}{\pi}\omega_{pd}^2 e^{-\kappa}(1+\kappa)\sin^2\left(\frac{qb}{2}\right) . \qquad (7.16)$$

One can see that the influence of the vertical confinement ω_0^2 is necessary to yield a stable dispersion relation.[2] The calculated dispersion relation is shown in Fig. 7.14. It is interesting to note that this wave is a backward wave ($\partial\omega/\partial q < 0$), that is phase

---

[2]Otherwise $\omega^2 \propto -\omega_{pd}^2\sin^2(qb/2)$ yielding a positive imaginary value for ω which indicates exponential growth.

Fig. 7.14 Dispersion of the transverse dust lattice wave for different values of κ without damping. The frequency scale, here, is normalized to ω_0, not ω_{pd}. Also note the limited frequency scale

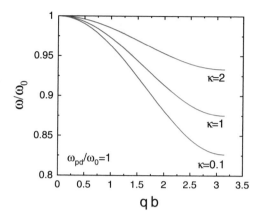

and group velocities move in opposite directions. In addition, the wave frequency approaches a finite value for long wavelengths ($\omega \rightarrow \omega_0$ for $q \rightarrow 0$). Such a type of wave is called "optical wave" in analogy to waves known from condensed matter. For $q \rightarrow 0$ this is a synchronous oscillation of all particles at the resonance frequency of the vertical potential well used for the determination of the dust charge (see Sect. 3.8).

Transverse dust lattice waves have been observed by Misawa et al. [19]. In their experiment, a linear chain of dust particles shows vertical oscillations (see Fig. 7.15) which propagate along the chain. From the time traces it is immediately seen that the wave is a backward wave (negative slope in the space-time diagram). The authors have measured a part of the dispersion relation where a finite frequency is found for $q \rightarrow 0$ and the dispersion also has a negative slope, as expected for the transverse DLW. However, the overall agreement of the measured and the theoretical dispersion was not very satisfying.

In a different experiment [9], this optical mode has been measured from laser excitation in a chain of particles with 8.09 μm diameter. There, both the real and imaginary part of the wave vector have been measured with very good agreement with the expected dispersion relation. This has been later refined by analyzing the natural modes [20], as shown in the next section.

7.8   Dispersion Relation from Thermal Particle Motion

In the past years, a method was developed by Nunomura et al. to determine the dispersion relation of the various wave modes from the thermal Brownian motion of the dust particles in the plasma crystal [8, 9, 20]. This powerful technique allows the measurement of the *entire* dispersion from a single video sequence without additional particle excitation.

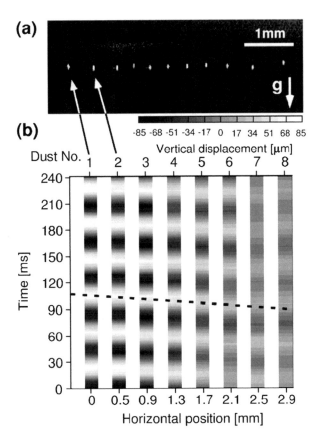

Fig. 7.15 Transverse dust lattice waves. (a) Still image of a 1D particle chain, (b) Grey scale image of the vertical displacement of the dust particles in the chain. The wave is seen to propagate backwards. Reprinted figure with permission from [19]. © (2001) by the American Physical Society

There, the thermal Brownian motions of the particles are measured and analyzed. Hereby it is assumed that the Brownian motion of the particles is an incoherent superposition of all wave modes ("phonons") of the crystal. Then, to extract the dispersion relation from the Brownian motion, first the particle velocities $\mathbf{v}(\mathbf{r}, t)$ in the 2D particle plane are determined. From that the Fourier components of the velocity

$$\hat{v}(\mathbf{q}, \omega) = \frac{2}{TL} \int_0^T \int_0^L \mathbf{v}(\mathbf{r}, t) e^{(-i\mathbf{q}\cdot\mathbf{r}+i\omega t)} \mathrm{d}\mathbf{r}\,\mathrm{d}t \qquad (7.17)$$

are determined. Here, L and T are the length and period over which the particle motion is integrated. The compressional mode is then obtained by taking the components of $\mathbf{v} \| \mathbf{q}$, only. The shear mode is derived from the components of $\mathbf{v} \perp \mathbf{q}$ in the plane and the transverse mode from the out-of-plane velocity components $\mathbf{v} \perp \mathbf{q}$. This integral is evaluated for all wave vectors $\mathbf{q}$ and frequencies ω. The square of $\hat{v}$ is then proportional to the wave intensity at the chosen values of $\mathbf{q}$ and ω.

Figure 7.16a, b shows the wave energy density of the Fourier components as a function of q and ω for the compressional and shear mode of an actual experiment with micron-sized particles [8, 9, 20]. Following the above procedure the wave energy is large along distinct lines indicating the dispersion relation of the two wave modes. The dispersion relations obtained from this Brownian motion technique are in very good agreement with the theoretical predictions. The compressional mode shows the expected sin-like behavior with the acoustic dispersion at long wavelengths and the bending-over at higher wave numbers. The shear mode exhibits the linear dispersion of the essentially acoustic behavior of this mode. This demonstrates the ability of this method to derive the dispersion relation from a single video sequence.

In another experiment, this method has also been applied to a linear chain of particles to measure the transverse mode dispersion (Fig. 7.16c). Again good agreement with the theoretical dispersion is obtained [9, 20]. This powerful technique has also been applied to finite systems in the following chapter where also some more details of this method are discussed.

7.9 A Note on Shielding

At this point, a few words on the screening of dust particles in the sheath of rf discharges are appropriate. We already briefly addressed the question of shielding in Sects. 3.3.2 and 4.1. The above mentioned wave experiments (as well as the experiments on finite clusters in the following chapter) have revealed that the screening strength $\kappa = b/\lambda_D$ is of the order of 1. This means that the shielding length λ_D is of the order of the interparticle distance b. The interparticle distance and thus the screening length is typically a few hundred microns for the experiments with micron-sized particles. As examples, $\lambda_D = 850\,\mu m$ is derived from the 1D experiment in Sect. 7.3 [4] and $\lambda_D = 750\,\mu m$ to $950\,\mu m$ are determined from the 2D experiments under different discharge conditions [7]. Direct collision experiments in Sect. 4.1 also have found screening lengths of the same size, namely $\lambda_D = 340\,\mu m$ and $780\,\mu m$ at different plasma conditions [21].

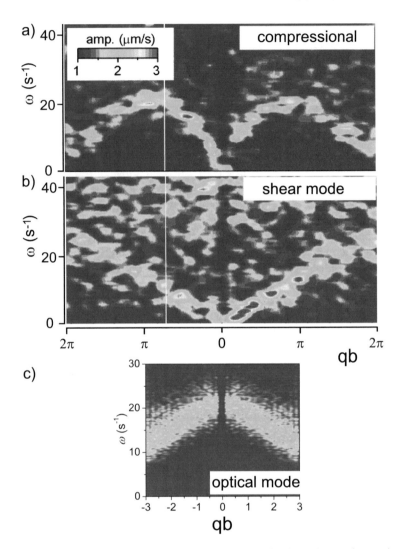

Fig. 7.16 Dispersion relation of the modes determined from the thermal Brownian motion. (**a**) compressional wave, (**b**) shear wave, (**c**) optical wave. Note that in these figures the wave vector extends from $qb = -2\pi$ to $qb = +2\pi$, thus $qb = 0$ is in the center. Reprinted figures with permission from [8, 20]. © (2002,2005) by the American Physical Society

A dust system where the shielding length is much shorter than the interparticle distance $\lambda_D \ll b$ the particles can be considered to behave like isolated, single particles, a situation sometimes referred to as "dust-in-plasma". The opposite situation with an interparticle distance much shorter than the shielding length ($\lambda_D \gg b$) is similar to an ideal plasma state where there are many dust particles within the shielding cloud. This then would then constitute an "ideal" dusty plasma.

As we now know, dusty plasmas in experiments are found to be just in between these two situations, namely at $\lambda_D \approx b$ ($\kappa \approx 1$).

We now would like to discuss these observed screening lengths in some more detail. The observed shielding lengths $\lambda_{D,obs}$ of some hundred microns are close to the electron Debye length

$$\lambda_{D,obs} \approx \lambda_{D,e} = \sqrt{\frac{\varepsilon_0 k_B T_e}{n_e e^2}} .$$

For typical plasma conditions of n_e around $10^8\,cm^{-3}$ and T_e around $5\,eV$ the electron Debye length is obtained as $\lambda_{D,e} \approx 550\,\mu m$. In comparison, the ion Debye length

$$\lambda_{D,i} = \sqrt{\frac{\varepsilon_0 k_B T_i}{n_i e^2}}$$

is about $60\,\mu m$ due to the much smaller ion temperature $T_i \approx 0.05\,eV$ in typical gas discharges. The combined action of electron and ion screening happens at the linearized Debye length which is given by, repeating Eq. (4.2),

$$\frac{1}{\lambda_D^2} = \frac{1}{\lambda_{D,i}^2} + \frac{1}{\lambda_{D,e}^2} .$$

Since the shorter length scale determines the screening length the linearized Debye length is very close to the ion Debye length, i.e. $\lambda_D \approx \lambda_{D,i} \ll \lambda_{D,obs}$. The observed screening length should be of the order of the linearized or ion Debye length if the ions would be responsible for screening. Moreover, ions would be the natural choice for the shielding species since their charge sign is opposite to that of the dust.

This apparent paradox that the observed shielding length is close to the electron Debye length is resolved by accounting for the ion streaming motion. The experiments presented here have all been performed in the sheath of a plasma discharge. The ions enter the sheath with Bohm velocity $v_B = (k_B T_e/m_i)^{1/2}$ and are further accelerated by the sheath electric field towards the electrode. Hence, one could argue that these ions at supersonic speeds $u_i \geq v_B$ cannot participate in screening: at supersonic speeds the ion motion is dominated by the streaming motion and not by the dust potential.

Khrapak et al. and Hutchinson [22, 23] have proposed an expression for the velocity dependent shielding length $\lambda_{D,eff}$ which describes the shielding properties of a plasma with ions drifting at a velocity u_i, compare Eq. (3.12). It is given as

$$\lambda_{D,eff}^2 = \frac{\lambda_{D,e}^2}{1 + k_B T_e/(k_B T_i + (1/2)m_i u_i^2)} , \tag{7.18}$$

Fig. 7.17 Screening length
$\lambda_{D,eff}$ (in units of the electron
Debye length $\lambda_{D,e}$) as a
function of ion drift velocity
u_i (in units of the ion thermal
velocity)

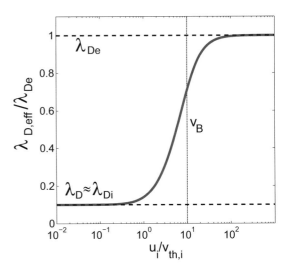

which is also shown in Fig. 7.17. For low drift velocities $u_i \to 0$ the screening
length is the linearized Debye length $\lambda_{D,eff} = \lambda_D$ whereas for high drift velocities it
approaches the electron Debye length $\lambda_{D,eff} = \lambda_{D,e}$. Hence, when the ion streaming
velocity is below the ion thermal velocity the appropriate screening length is the
linearized Debye length. For large drifts $u_i \gg v_{th,i}$ the appropriate screening length
is the electron Debye length $\lambda_{D,e}$. Since the experiments are performed in the sheath
where $u_i \geq v_B \gg v_{th,i}$ screening at the electron length scale is dominating.

References

1. F. Melandsø, Phys. Plasmas **3**, 3890 (1996)
2. F.M. Peeters, X. Wu, Phys. Rev. A **35**, 3109 (1987)
3. X. Wang, A. Bhattacharjee, S. Hu, Phys. Rev. Lett. **86**, 2569 (2001)
4. A. Homann, A. Melzer, S. Peters, R. Madani, A. Piel, Phys. Rev. E **56**, 7138 (1997)
5. A. Homann, A. Melzer, R. Madani, A. Piel, Phys. Lett. A **242**, 173 (1998)
6. S. Nunomura, D. Samsonov, J. Goree, Phys. Rev. Lett. **84**, 5141 (2000)
7. S. Nunomura, J. Goree, S. Hu, X. Wang, A. Bhattacharjee, Phys. Rev. E **65**, 066402 (2002)
8. S. Nunomura, J. Goree, S. Hu, X. Wang, A. Bhattacharjee, K. Avinash, Phys. Rev. Lett. **89**, 035001 (2002)
9. B. Liu, K. Avinash, J. Goree, Phys. Rev. Lett. **91**, 255003 (2003)
10. D. Samsonov, J. Goree, Z. Ma, A. Bhattacharjee, H. Thomas, G. Morfill, Phys. Rev. Lett. **83**, 3649 (1999)
11. D. Samsonov, J. Goree, H. Thomas, G. Morfill, Phys. Rev. E **61**, 5557 (2000)
12. A. Melzer, S. Nunomura, D. Samsonov, J. Goree, Phys. Rev. E **62**, 4162 (2000)
13. D.H.E. Dubin, Phys. Plasmas **7**, 3895 (2000)
14. V. Nosenko, J. Goree, Z.W. Ma, A. Piel, Phys. Rev. Lett. **88**, 135001 (2002)
15. O. Havnes, T. Aslaksen, T.W. Hartquist, F. Li, F. Melandsø, G.E. Morfill, T. Nitter, J. Geophys. Res. **100**, 1731 (1995)

16. K. Jiang, V. Nosenko, Y.F. Li, M. Schwabe, U. Konopka, A.V. Ivlev, V.E. Fortov, V.I. Molotkov, A.M. Lipaev, O.F. Petrov, M.V. Turin, H.M. Thomas, G.E. Morfill, EPL (Europhys. Lett.) **85**(4), 45002 (2009). http://stacks.iop.org/0295-5075/85/i=4/a=45002
17. D.I. Zhukhovitskii, V.E. Fortov, V.I. Molotkov, A.M. Lipaev, V.N. Naumkin, H.M. Thomas, A.V. Ivlev, M. Schwabe, G.E. Morfill, Phys. Plasmas (1994–present) **22**, 023701 (2015). http://scitation.aip.org/content/aip/journal/pop/22/2/10.1063/1.4907221
18. S.V. Vladimirov, P.V. Shevchenko, N.F. Cramer, Phys. Rev. E **56**, 74 (1997)
19. T. Misawa, N. Ohno, K. Asono, M. Sawai, S. Takamura, P. Kaw, Phys. Rev. Lett. **86**, 1219 (2001)
20. B. Liu, J. Goree, Phys. Rev. E **71**, 046410 (2005)
21. U. Konopka, G. Morfill, L. Ratke, Phys. Rev. Lett. **84**, 891 (2000)
22. S.A. Khrapak, A.V. Ivlev, S.K. Zhdanov, G.E. Morfill, Phys. Plasmas **12**, 042308 (2005)
23. I.H. Hutchinson, Plasma Phys. Control. Fusion **48**, 185 (2006)

Chapter 8
Finite Dust Clusters

So far, waves in extended (1D and 2D) systems as well as crystalline structures in extended 2D and 3D systems have been presented. Now, we would like to draw the attention to collective effects in finite systems. Such Coulomb or Yukawa clusters are systems of a small number of charged particles trapped in an external potential and interacting by mutual Coulomb/Yukawa repulsion. As early as 1904, such systems have been analyzed by Thomson as a model for the structure of the atom [1]. Finite clusters are particularly appealing because of the interplay between the electrostatic repulsion among the particles and the confinement due to an external potential. This interplay determines the structure of a cluster and also its dynamics. Structure and dynamics dramatically depend on the exact particle number N, and the formation of highly symmetric configurations with "magic" particle numbers is observed in a wide variety of situations in the classical and quantum world.

Examples of charged-particle clusters of present-day interest are found as ions in Paul and Penning traps [2], ions in storage rings of accelerators [3], electrons on the surface of liquid helium [4], electrons in quantum dots [5], brown dwarf stars [6], charged particles in colloidal suspensions [7] or, as will be discussed here, charged dust particles in dusty plasmas [8].

8.1 Formation of Finite Dust Clusters

Finite dust clusters are formed by trapping only a small number of dust particles into a confinement potential provided by a combination of forces described in Chap. 3.

© Springer Nature Switzerland AG 2019
A. Melzer, *Physics of Dusty Plasmas*, Lecture Notes in Physics 962,
https://doi.org/10.1007/978-3-030-20260-6_8

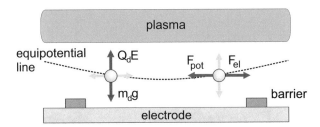

Fig. 8.1 Sketch of the confinement of clusters in an RF discharge. In the vertical direction a very strong confinement is provided by the electric field force and gravity. In the horizontal direction a barrier on the electrode provides a shallow horizontal confinement. The force due to that confinement potential F_{pot} balances the electrostatic repulsion F_{el} of the particles

The confinement potential energy V depends on the peculiarities of the setup and the exploited forces (see Fig. 8.1), but can often be considered as harmonic, i.e.

$$V(x_i, y_i, z_i) = \frac{1}{2}m_d\omega_x^2 x_i^2 + \frac{1}{2}m_d\omega_y^2 y_i^2 + \frac{1}{2}m_d\omega_z^2 z_i^2$$

$$= \frac{1}{2}m_d\omega_0^2 \left(x_i^2 + \alpha_y y_i^2 + \alpha_z z_i^2 \right). \tag{8.1}$$

Here, $\omega_{x,y,z}$ is the confinement strength in x, y, z direction and $\alpha_{y,z} = \omega_{y,z}^2/\omega_x^2$ is the relative strength of the confinement with respect to the confinement in x direction ($\omega_0 = \omega_x$). Further, x_i, y_i, z_i denote the position of particle i along the spatial axes. By changing the relative confinement strengths $\alpha_{x,y}$ various confinement geometries from 1D to 2D and 3D can be realized. How this is done in the experiment will be demonstrated below.

8.2   Structure of 1D Dust Clusters

Linear (1D) dust clusters are easily generated by placing a rectangular metal barrier onto the lower electrode (see Fig. 8.2 or [9]) or by cutting a groove into the electrode [10]. A small number of particles ($N = 1$ to 20) are then dropped into the space charge sheath above the barrier [11].

Vertically the particles are strongly confined due to the balance of electric field force and gravity (i.e. $\omega_z \gg \omega_{x,y}$ or $\alpha_z \gg 1$). Horizontally the particles are confined due to the barrier on the electrode. Since the elongation of the barrier is much larger in y-direction than in x-direction the particles are only weakly confined along y. Thus, $\alpha_y \ll 1$ and the particles arrange along y in a linear 1D dust cluster.

Figure 8.2b shows the arrangement of $N = 4$ to $N = 18$ particles in the barrier. It is seen that for $N = 4$ to $N = 9$ the particles strictly arrange in a linear arrangement. When, however, the particle number is increased from 9 to 10 a zigzag transition in

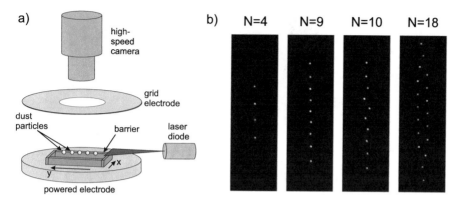

Fig. 8.2 (**a**) Scheme of the experimental setup for the confinement of 1D dust clusters. In this particular case, the barrier is 6 mm high and 5 × 40 mm wide. (**b**) Snap shots of the 1D dust clusters for $N = 4, 9, 10$ and 18 particles of 10.2 μm diameter. A structural transition in the cluster is seen by increasing the particle number. From 9 to 10 particles a zigzag transition occurs. From [11]

the center of the chain occurs. For 18 particles a zigzag structure is seen nearly throughout the entire chain.

The reason for that is easily understandable: the confinement in the y-direction compresses the chain along its extension. The interparticle distance is smallest in the central part of the chain. When more and more particles are inserted into the chain the compression increases until it is easier for the central particles to make a transverse excursion (in the x-direction). Then, the force along y between neighboring particles due to compression exceeds the force in x from the confinement [11], see also [12].

8.3 Structure of 2D Finite Dust Clusters

In 2D clusters, a small number of dust particles $N = 1$ to 1000, say, is trapped in the sheath above the lower electrode [13–15]. Vertically, the particles are, as usual, strongly confined by the electric field force and gravity. In the horizontal plane, now a weak circular parabolic confinement (of strength ω_0) is applied. Such a confinement can be realized, e.g., by a shallow circular parabolic trough in the electrode (see Fig. 8.3). Hence, here, $\omega_0^2 = \omega_x^2 = \omega_y^2 \ll \omega_z^2$ ($\alpha_y = 1, \alpha_z \gg 1$).

Under the interplay of the horizontal parabolic confinement and their electrostatic repulsion the particles arrange in concentric shells (see Fig. 8.4). The 2D clusters are sometimes said to form a "periodic table". The structure and their dynamic properties dramatically depend on the particle number N. When changing from 5 to 6 particles (or from 15 to 16) a new shell opens up. Like in atomic and nuclear physics there exist "magic" particle numbers of high dynamic stability, e.g. the

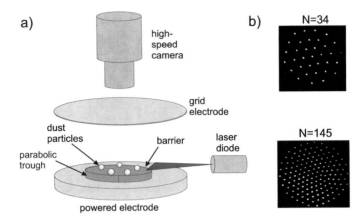

Fig. 8.3 (**a**) Scheme of the experimental setup for the confinement of 2D dust clusters using a parabolic trough. (**b**) Snap shots of the 2D dust clusters for $N = 34$ and 145 particles of 9.55 μm diameter. From [16]

$N = 19(1, 6, 12)$ cluster. The notation $(N_1, N_2, N_3, \ldots)$ refers to N_1 particles in the inner ring, N_2 in the second and so on. These clusters will be analyzed below in more detail.

These finite 2D clusters can be described in terms of their total energy

$$E = \frac{1}{2} m_d \omega_0^2 \sum_{i=1}^{N} r_i^2 + \frac{Z_d^2 e^2}{4\pi \varepsilon_0} \sum_{i>j}^{N} \frac{\exp(-r_{ij}/\lambda_D)}{r_{ij}}, \qquad (8.2)$$

where $\mathbf{r}_i = (x_i, y_i)$ is the 2D position of the i-th particle in the horizontal plane and $r_i = |\mathbf{r}_i|$ is the radial distance from the trap center. Then, $r_{ij} = |\mathbf{r}_i - \mathbf{r}_j|$ is the distance between particles i and j. The strength of the horizontal confinement is denoted by the horizontal resonance frequency ω_0. The first term is the potential energy due to the confinement in the horizontal plane and the second is the Coulomb/Yukawa repulsion of the particles.

When measuring the distances $(r \to r/r_0)$ and energies $(E \to E/E_0)$ in units of

$$r_0 = \left[\frac{Z_d^2 e^2}{4\pi \varepsilon_0} \frac{2}{m_d \omega_0^2} \right]^{1/3} \quad \text{and} \quad E_0 = \left[\left(\frac{Z_d^2 e^2}{4\pi \varepsilon_0} \right)^2 \frac{m_d \omega_0^2}{2} \right]^{1/3} \qquad (8.3)$$

the energy relation reduces to

$$E = \sum_{i=1}^{N} r_i^2 + \sum_{i>j}^{N} \frac{\exp(-\kappa r_{ij})}{r_{ij}}, \qquad (8.4)$$

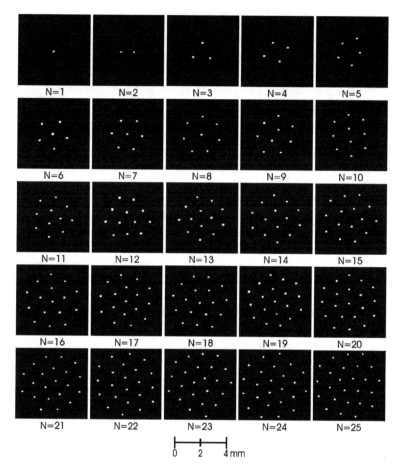

Fig. 8.4 "Periodic table" of finite 2D clusters with $N = 1$ to $N = 25$ particles of 9.55 μm diameter. These images are taken from the experiment [16]

where $\kappa = r_0/\lambda_D$ is the screening strength for finite systems. The parameter r_0 is a measure for the interparticle distance. In these normalized units the energy and the distances do not depend on the physical parameters like particle charge, mass etc. They only define the absolute values of the energy and the absolute size of the cluster, but they do not have an influence on the structure, on the dynamics and so on. Structure and dynamics are fully determined by the particle number N and the screening strength κ.

The equilibrium structure of these systems is derived from the minimum of the total energy [17]. Experimentally they have been observed, e.g. by Juan et al. [13], Klindworth et al. [14] or Cheung et al. [15]. The observed cluster structures are generally in perfect agreement with the theoretical predictions.

8.4   Structure of 3D Finite Dust Clusters

It is also possible to confine 3D dust clusters under laboratory conditions. The challenge is to produce an isotropic confinement in 3D that does not lead to void formation. The confinement has been achieved using the combined interaction of thermophoretic levitation and a horizontal boundary due to glass walls, see Fig. 8.5 and [18, 20, 21]. There, a glass box is placed onto the lower electrode. The dielectric walls charge negatively and therefore provide an inward electric force on the negative dust particles for horizontal confinement. The lower electrode is heated and induces a temperature gradient in the neutral gas. The temperature gradient provides an upward thermophoretic force that, at least partially, compensates the gravitational force on the particles [18]. By tuning the electrode temperature and plasma properties a 3D isotropic confinement potential is formed to confine the dust particles ($\omega_x^2 = \omega_y^2 = \omega_z^2$, i.e. $\alpha_y = \alpha_z = 1$).

The clusters in this three-dimensional confinement also arrange in highly ordered structures [22]. The systems form concentric spheres, arranged in an "onion shell" structure, for which the name "Yukawa ball" has been established [19, 20, 23, 24]. Like 2D finite clusters where the particles arrange in concentric rings, 3D clusters have clearly separated spheres. Since a 3D structure is difficult to visualize, the structure is most clearly seen in cylindrical coordinates (ρ, z), where the horizontal coordinates (x, y) are mapped onto the radial position $\rho = \sqrt{x^2 + y^2}$ and plotted versus the vertical coordinate z. An experimental dust cluster is shown in Fig. 8.6. The cluster consists of $N = 190$ particles and is arranged in 4 concentric shells in a (2, 21, 60, 107) configuration, i.e. 2 particles are on the innermost shell, followed by 21, 60 and 107 particles on the second, third and forth shell, respectively [20]. The structure within the two outer shells is visualized in Fig. 8.6b, c where a Wigner–Seitz-cell construction has been performed on the individual shells. The shells

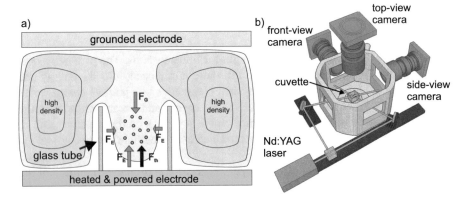

Fig. 8.5 (**a**) Scheme of the experimental setup for the confinement of 3D dust clusters. Thermophoretic levitation is used to (partially) compensate the gravitational force. (**b**) Scheme of the 3D stereoscopic imaging unit. From [18, 19]

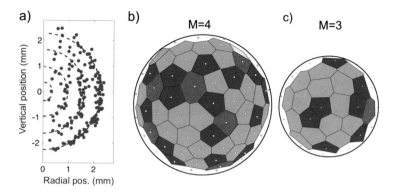

Fig. 8.6 3D dust cluster in an experiment with $N = 190$ particles of $3.4\,\mu m$ diameter. (**a**) Particle positions in cylindrical $\rho - z$ coordinates. The cluster is seen to have 4 concentric shells. (**b**) Wigner–Seitz-cell analysis of the outer (4th) shell and (**c**) of the 3rd shell. Pentagons are marked blue, hexagons are green. Defect polygons with more than 6 nearest neighbors are colored red. From [20]

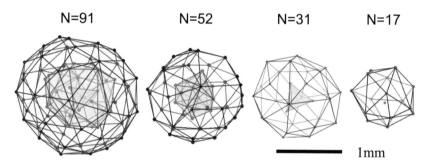

Fig. 8.7 Cluster configurations reconstructed from single video snap shots with $N = 91, 52, 31$, and 17 particles of $3.47\,\mu m$ diameter. The Yukawa balls consist of concentric shells with the configurations $(4, 25, 62)$ for the $N = 91$, $(11, 41)$ for the $N = 52$, $(5, 26)$ for the $N = 31$, and $(1, 16)$ for the $N = 17$ cluster. From [19]

consist of hexagons and pentagons (like the famous C_{60} buckyballs or old-fashioned soccer balls). A certain number of pentagons is needed to ensure the curvature of the sphere and to form closed shells. But also some defects, i.e. particles with more than 6 nearest neighbors are found.

A selection of smaller 3D clusters ($N < 100$) is shown in Fig. 8.7. There, a 3D reconstruction of the structure is shown. Also these clusters arrange in nested spherical shells ranging from one shell with central particle ($N = 17$) to three shells ($N = 91$). The highly ordered arrangement of the particles in the cluster is readily seen. Analogously to the 2D case, also here a "periodic table"-like construction of the clusters is observed. Typically, a new shell opens up, when the inner shell has 12 particles [25, 26].

However, there is an interesting difference between 2D and 3D clusters concerning the structural properties and screened particle interaction. In 2D, the observed cluster structures (i.e. occupation numbers of the different rings) is nearly independent of the particle interaction: pure Coulomb, Yukawa or even logarithmic interaction potentials (almost) always yield the same cluster structure [27]. In contrast, in 3D, for fixed total particle number N the occupation number of the different shells is different between Yukawa interaction and pure Coulomb interaction. For Yukawa interaction, the occupation number of inner shells is higher than for the Coulomb case. Consequently, outer shells have lower particle numbers for Yukawa interaction than for Coulomb interaction. Hence, already from the structure of 3D clusters the shielding strength can be extracted [19, 23, 24]. From the occupation numbers of the experimental clusters a screening strength of, again, $\kappa \approx 1$ has been deduced [23]. These non-negligible screening effects justify the term "Yukawa ball" for these clusters.

The structural difference between 3D finite Coulomb and Yukawa systems can also be modeled from continuum models that provide enlightening insights [28–30]. Considering a shell of the dust cluster as a homogeneously charged sphere, then in the case of pure Coulomb interaction another charge inside this sphere would not feel a force, since the electric field inside the charged sphere is zero. In the case of Yukawa interaction, a force towards the center of the shell remains (since the charge contributions from the opposite side of the shell are reduced due to shielding). This force towards the center has to be counter-balanced by more particles on the inside leading to the observed structures with larger population on the inner shells.

8.5   Normal Modes in Finite Clusters

After the discussion of the static equilibrium properties of finite systems we like to delve into the dynamic properties of these systems. The dynamic properties of finite systems cannot be described in terms of pure compressional or shear waves as in the case of extended systems. First, only a certain number of wavelengths can fit into finite systems. Moreover, due to the boundaries of finite systems, where phase jumps and reflections can occur, the distinction between shear and compressional modes is not possible.

The dynamic properties of finite clusters are described by their normal modes which replace the dispersion relation of infinite systems [31]. These modes in finite systems have both compressional and shear contributions. For small systems the boundaries are important and dominate the dynamic properties of the system. We will treat the problem for the 2D case, here, for simpler visualization. The extension to 3D systems is easily done.

The starting point is the total energy E from Eq. (8.4). The absolute value of E is determined by the ground state configuration of the cluster. Since the ground state is an equilibrium, we have

$$
\begin{pmatrix}
\partial/\partial x_1 \\
\vdots \\
\partial/\partial x_N \\
\partial/\partial y_1 \\
\vdots \\
\partial/\partial y_N
\end{pmatrix}
E = 0,
$$

where the derivatives along x and y direction for all particles $i = 1$ to N vanish. The mode oscillations are then determined from the harmonic (parabolic) expansion around the equilibrium positions.

For multi-particle systems it is known from theoretical mechanics that this harmonic expansion is obtained from the $2N \times 2N$ dynamical matrix (for 2D systems)

$$
\mathsf{A} =
\begin{pmatrix}
\left(\dfrac{\partial^2 E}{\partial x_i \partial x_j} \right) & \left(\dfrac{\partial^2 E}{\partial x_i \partial y_j} \right) \\[3ex]
\left(\dfrac{\partial^2 E}{\partial y_i \partial x_j} \right) & \left(\dfrac{\partial^2 E}{\partial y_i \partial y_j} \right)
\end{pmatrix},
\tag{8.5}
$$

that contains the second derivatives of the total energy with respect to all particle coordinates x_i and y_j. These second derivatives are similar to that in Eq. (7.10) where we have seen in the determination of the DLW that the effective "spring constant" of the interaction among the different particles is related to the second derivative of the interaction potential. The dynamical matrix A is just the generalization of this concept for multi-particle systems in more than one dimension.

In extended form, the 4 elements that constitute the dynamical matrix A are themselves $1N \times 1N$ matrices that contain the possible combinations of i and j, namely, e.g.

$$
\left(\frac{\partial^2 E}{\partial x_i \partial y_j} \right) =
\begin{pmatrix}
\dfrac{\partial^2 E}{\partial x_1 \partial y_1} & \cdots & \dfrac{\partial^2 E}{\partial x_1 \partial y_N} \\[3ex]
\vdots & \ddots & \vdots \\[2ex]
\dfrac{\partial^2 E}{\partial x_N \partial y_1} & \cdots & \dfrac{\partial^2 E}{\partial x_N \partial y_N}
\end{pmatrix}.
$$

To obtain the normal modes of the N-particle system the eigen value problem

$$
A \begin{pmatrix} x_1 \\ \vdots \\ x_N \\ y_1 \\ \vdots \\ y_N \end{pmatrix} = \omega^2 \begin{pmatrix} x_1 \\ \vdots \\ x_N \\ y_1 \\ \vdots \\ y_N \end{pmatrix} \tag{8.6}
$$

has to be solved.[1]

The eigen values and eigen vectors of A describe the normal mode oscillations of the finite clusters. The eigen values ω^2 are the oscillation frequencies and the eigen vectors $(x_1, \ldots, x_N, y_1, \ldots, y_N)^T$ describe the mode oscillation patterns. The matrix A is $2N \times 2N$, thus there are $2N$ eigen modes for a system with N particles in two dimensions (consequently, in 3D, the matrix A is $3N \times 3N$ with $3N$ eigen modes).

This is demonstrated for the simple case of $N = 3$ particles in 2D in Fig. 8.8 where the $2N$ eigen modes of the cluster are presented. There are modes that are common to all clusters. These common modes are, first, the two sloshing modes, i.e. oscillations of the entire cluster in the horizontal confining potential in x and in y direction, modes number 5 and 6 (in 3D there are, of course, 3 sloshing modes). Then, there is the rotation of the entire cluster (mode number 2) and the breathing mode, i.e. a coherent, purely radial oscillation of all particles, mode number 1. It is easily seen that the mode frequency for the two sloshing modes (5, 6) is $\omega_{sl} = \omega_0$ since the cluster oscillates as a whole in the confinement potential of strength ω_0. For the rotation (2) of the entire cluster $\omega_{rot} = 0$ since there are no restoring forces and the cluster is in equilibrium for any rotation angle.

The breathing mode (1) has the interesting property that for pure Coulomb interaction, i.e. $\kappa = 0$, its frequency always is $\omega_{br} = \sqrt{3}\omega_0$ independent of particle number [31]. For shielded interaction ($\kappa > 0$) the frequency of the breathing mode increases and slightly depends on particle number N (see Fig. 8.9).

For the 3-particle cluster also two "kink" modes are found (mode number 3 and 4). These modes are specific for the $N = 3$ cluster and describe an oscillatory motion where at a certain instant two particles approach each other and the third moves away. This mode frequency is found at $\omega_k = \sqrt{3/2}\omega_0$ for $\kappa = 0$ and increases with κ. Of course, the frequencies of the sloshing mode and cluster

---

[1] This is easily seen from a simple analogy: Starting from the equation of motion for a simple spring

$$
m\ddot{x} = -kx \quad \rightarrow \quad -\omega^2 x = -(k/m)x
$$

assuming an oscillatory solution $x \rightarrow x \exp(-i\omega t)$. Now, for the many-particle case k/m is replaced by the second derivatives, i.e the dynamical matrix A, and x becomes the vector of particle positions in x, y.

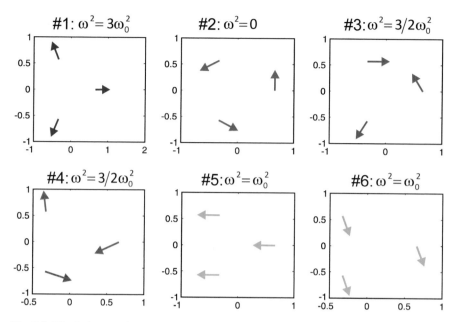

Fig. 8.8 The 6 eigen modes of a $N = 3$ cluster in 2D with the corresponding mode frequencies ω^2 (for $\kappa = 0$)

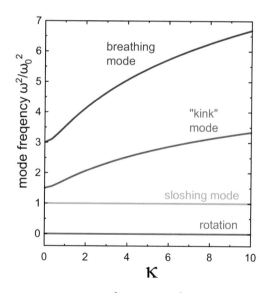

Fig. 8.9 Evolution of the mode frequency ω^2 (in units of ω_0^2) of the 6 eigen modes of the 3 particle cluster. From [16]

rotation are independent of κ and stay at their frequencies $\omega_{sl} = \omega_0$ and $\omega_{rot} = 0$, respectively, as shown in Fig. 8.9.

8.6 Modes from Thermal Particle Motion

As for the case of the waves, the normal modes can be extracted experimentally from the thermal Brownian motion of the dust particles [16]. Likewise, it is assumed that the Brownian motion of the particles is a superposition of all eigen modes of the cluster. As for the wave analysis, first, also for the finite clusters the velocity of the Brownian motion of all particles $\mathbf{v}_i(t)$ is determined from a video sequence. Then, the contribution of the thermal motion to each of the eigen modes is determined from the projection of the thermal velocities onto the eigen mode pattern by

$$f_\ell(t) = \sum_{i=1}^{N} \mathbf{v}_i(t) \cdot \mathbf{e}_{i,\ell},$$

where $\mathbf{e}_{i,\ell}$ is the eigen vector for particle i in mode number ℓ, as determined from Eq. (8.6). The function $f_\ell(t)$ is the contribution of the thermal motion to the eigen mode ℓ in the time domain. In comparison, in the wave analysis (Sect. 7.8), the Brownian particle velocities were projected onto the chosen wave vector $\mathbf{q}$. For the clusters, the eigen mode vectors take the role of the wave vectors.

Finally, the spectral power density of each mode ℓ

$$S_\ell(\omega) = \frac{2}{T} \left| \int_0^T f_\ell(t) e^{i\omega t} dt \right|^2 \tag{8.7}$$

of f_ℓ is calculated. The spectral power density is the square of the Fourier transform of $f_\ell(t)$. The power spectrum contains the contribution of the thermal motion to each of the $2N$ eigen modes in the frequency domain and is proportional to the energy stored in the modes. It can be shown that

$$\int_0^\infty S_\ell(\omega) \, d\omega = \langle v_\ell^2 \rangle$$

thus from the equipartition theorem

$$\frac{1}{2} m_d \int_0^\infty S_\ell(\omega) \, d\omega = \frac{1}{2} m_d \langle v_\ell^2 \rangle = \frac{1}{2} k_B T_\ell = E_\ell$$

is the energy E_ℓ or the temperature T_ℓ stored in mode number ℓ.

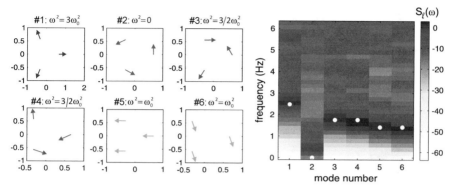

Fig. 8.10 Spectral power density of the 3 particle cluster derived from the thermal motion of the particles. The white dots indicate the best-fit theoretical values of the mode frequencies. The 6 eigen modes of a $N = 3$ cluster with the corresponding mode frequencies ω^2 (in units of ω_0^2) are also shown again for convenience. From [16]

In an experiment, a three-particle cluster has been trapped and the thermal motion has been measured. The power spectrum of this cluster has been obtained and is shown in Fig. 8.10. There, darker colors represent higher spectral power density. So, mode number 1 shows its main response at a frequency of about 2.5 Hz under these conditions, mode number 2 at a frequency of less than 1 Hz and so on. The measured spectrum is then also compared with the theoretical mode frequencies, i.e. the eigen values of the dynamical matrix. Since the theoretical mode frequencies of *all* $2N$ modes depend only on the two parameters ω_0 and κ, the theoretical mode frequencies were fitted to the experimental data by changing these two values. It is seen, that the observed power spectrum is in very good agreement with the expected mode frequencies.

This technique can be applied to clusters of any size and has been demonstrated for clusters with several hundred particles. From the analysis of the mode spectra the best-fit parameters ω_0 and κ are derived. Note again, that only these two parameters are fitted to match all $2N$ modes simultaneously, thus giving the obtained values a high credibility. Then, from the measured absolute values of cluster size or interparticle distance and by comparison with the scaling parameter $r_0 = [2Z_d^2 e^2/(4\pi\varepsilon_0 m_d\omega_0^2)]^{1/3}$ also the particle charge can be directly extracted. The particle charge was found to be in the presented experiments about 10,000 elementary charges on a $2a = 9.5\,\mu\text{m}$ particle and the screening strength κ was found in the range between 0.5 and 2 [16]. This is exactly in line with the previously discussed values for the charge and the screening strength.

8.7 Stability

The claim was made that the power spectrum contains the complete dynamic information on the cluster. This is illustrated here for the very basic dynamic property, the stability against perturbations.

In order to perturb a cluster configuration a force against the restoring forces of the cluster needs to be applied. If the forces needed to induce perturbations are large then the system is more stable than if only small forces are required. In other words, if the restoring forces of the cluster are large against an applied perturbation the cluster is stable. Since the restoring forces directly determine the frequencies of the various modes, the mode with the lowest eigen frequency is the one that determines the stability of the entire cluster. The easiest way to disturb the cluster is along the mode with the lowest eigen frequency.

The rotational mode always has an eigen frequency of $\omega = 0$ and thus is always the mode with the lowest eigen frequency. However, a rotation of the entire cluster does not change the cluster configuration and thus is not really a perturbation of the cluster. So, the rotation mode is excluded from our following analysis and we seek for the eigen mode with the lowest frequency besides cluster rotation.

As an example we take the 19-particle cluster (see inset in Fig. 8.11a), which is a "magic number" configuration (1, 6, 12) due to its hexagonal symmetry of inner and outer ring. The particle number in inner and outer ring is commensurable. The inner and outer ring are locked into each other like the teeth of a tooth-wheel. One would expect that this cluster configuration is very stable, justifying the notion of "magic number". The power spectrum to investigate the stability of the cluster is shown in Fig. 8.11b. One sees that there is a large gap between zero frequency (rotation) and the next lowest eigen frequency at about 0.9 Hz. This indeed quite large frequency gap shows that the lowest eigen mode already has a quite high frequency which means strong restoring forces and thus a high stability.

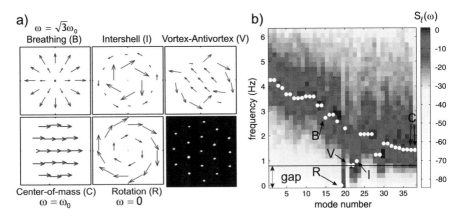

Fig. 8.11 (a) Selected modes of a cluster with $N = 19$ particles together with a video snap shot of that cluster. (b) Measured mode spectrum of this 19-particle cluster. From [16]

The corresponding mode patterns are shown in Fig. 8.11a. The already established breathing mode, center-of-mass mode and rotation are indicated. Two modes, that indicate the modes with the lowest frequency are the vortex-antivortex mode (V) and the intershell rotation (I). The vortex-antivortex mode consists of two vortices, a clockwise and a counter-clockwise rotation within the cluster. In this cluster, the vortex-antivortex mode has the lowest eigenfrequency (apart from rotation) and, thus, describes the pattern which results in the easiest disturbance of the 19-particle cluster. A similarly low frequency is found for the intershell rotation where we find a differential rotation of inner and outer rings. Vortex-antivortex formation and intershell rotation very often are the lowest frequency mode in 2D clusters.

For comparison, the mode spectrum of a $N = 20$ particle cluster is shown in Fig. 8.12. This cluster has a configuration of $(1, 7, 12)$ which is not a magic number configuration, since the number of particles in the inner and outer ring are not commensurable. Hence, inner and outer ring cannot interlock. Consequently, the intershell rotation is found at extremely low frequency of about 0.1 Hz, which is drastically less than for the 19-particle cluster. This demonstrates that the 20-particle cluster is very unstable against this intershell rotation. In contrast, the breathing mode of this cluster is found nearly at the same frequency as for $N = 19$ indicating that the plasma and confinement parameters of these two clusters are nearly identical. Also, the vortex-antivortex mode has nearly the same frequency as for the $N = 19$ cluster, and is not the lowest frequency mode.

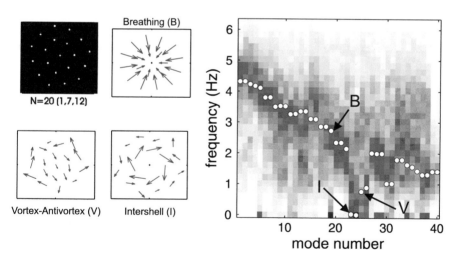

Fig. 8.12 Video snap shot of a cluster with $N = 20$ particles together with the measured mode spectrum. The breathing mode, the intershell rotation and the vortex-antivortex mode are marked by B, I and V, respectively. From [16]

8.8 Ion-Focus-Induced Phase Transitions in 2D Clusters

As another example of mode analysis in 2D clusters, the phase transition driven by
the ion-focus instability by reducing the gas pressure is revisited.

The mechanism of the phase transition has been introduced in Sect. 5.9. There,
we have seen there that the ion flow in the plasma sheath results in the formation
of positive ion space charges ("ion focus", "wakefield") beneath the dust particles
leading to the formation of the vertically aligned pairs of particles. Further, the
horizontal particle interaction becomes non-reciprocal due to the directed ion flow.
We have seen that, with reduced gas pressure, the vertical alignment becomes
unstable ("Schweigert instability") and gives rise to growing oscillations until a fluid
state is reached.

Here, we modify the situation in that we investigate a 2D finite system (with
about 40 particles of 9.55 μm diameter) instead of an extended system [32] and
only *one single* particle is placed in the layer below the actual cluster (see Fig. 8.13,
compare also Fig. 4.4). So there is only a single vertically aligned pair of particles.
The advantages of such a system are that, first, the heating effect can be definitely
attributed to the single lower-layer particle, and, second, the full dynamics in terms
of the normal modes of the cluster is accessible from the mode spectra (the mode
spectra are obtained only for the cluster of the upper particles, the lower particle is
a mere heat source).

As in the case with the extended system, the phase transition is induced by
reduction of gas pressure. With decreasing gas pressure, the single lower layer
particle starts to oscillate about its vertically aligned equilibrium position due to the
instability arising from the non-reciprocal attraction. Due to the mutual Coulomb
repulsion, the oscillating lower particle heats the upper-layer particles.

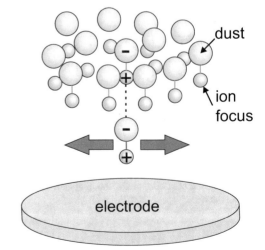

Fig. 8.13 Scheme of the instability-driven phase transition in finite 2D clusters. A single particle is confined below the actual cluster. The (negative) dust particles are shown together with their (positive) ion focus. By reducing the gas pressure oscillations of the lower particle are excited due to the Schweigert instability

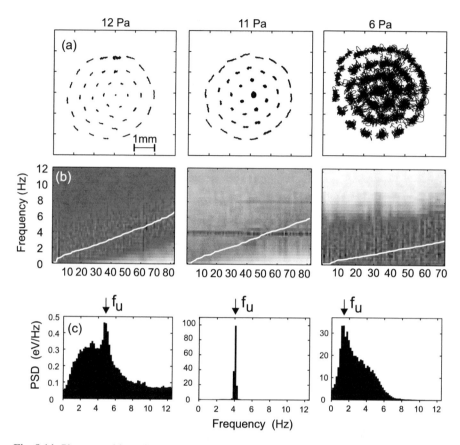

Fig. 8.14 Phase transition of a 2D cluster with one additional particle in the lower layer: (**a**) particle trajectories, (**b**) mode-resolved power spectra, (**c**) mode-integrated power spectra. From [32]

For illustration of the melting transition of the dust cluster, the particle trajectories are shown in Fig. 8.14a for different gas pressures. At the highest gas pressure (12 Pa) the particles only slightly move around their equilibrium positions, at reduced pressure (11 Pa) the oscillations become visible from the circular particle trajectories in the cluster center. At even lower pressure the particles start to exchange equilibrium positions which is an indication of melting, at 6 Pa frequent exchanges take place.

The dynamics of the cluster melting process is visible in great detail by analysis of the power spectra, see Fig. 8.14b. For 12 Pa, the spectral power density of the individual modes is concentrated around a quite narrow band of frequencies that closely follows the theoretical mode frequencies of a solid cluster with a particle charge of $Z_d = 11,000 \pm 1000$ and a screening length of $\lambda_D = (1000 \pm 500)\,\mu m$ (corresponding to $\kappa = 1 \pm 0.5$).

With reduced gas pressure (11 Pa), the spectrum changes completely. *All* modes show a maximum at the *same* frequency $f_u = 4$ Hz which is apparent from the dark horizontal band in the spectrum. This frequency of $f_u = 4$ Hz corresponds to the unstable oscillation of the lower layer particle. The dominance of this frequency in *all* modes is surprising. Nevertheless, apart from this dominant frequency, the underlying mode structure of the crystalline state is still faintly observable in the spectrum. From the mode spectra it is seen that the unstable oscillations appear in all the modes although only a single vertically aligned pair exists in this cluster. At 11 Pa the aligned pair starts the unstable oscillations where upper and lower particle vibrate around their equilibrium positions. Due to the Coulomb repulsion between the particles the oscillation is communicated to all particles in the upper layer and is thus visible in all the modes.

Below 8 Pa the situation changes again. The spectrum becomes broad for all modes and the close relation to the solid-state mode frequencies is lost. From this, it becomes obvious that the cluster is in a liquid state. It is interesting to note, here, that a single aligned pair is sufficient to drive the entire system of 40 particles into the liquid state. That means, that the oscillations of the vertical aligned pair "pump" so much energy into the system that the entire cluster melts in spite of the still relatively large frictional damping.

The analysis of the dynamic properties allows to fully substantiate the findings from the extended two-layer system. In this situation of the 2D cluster detailed information about the onset, the frequencies, the energies and the mechanisms of the melting transition are easily visible.

8.9   Laser-Heated Phase Transitions in 3D Clusters

Finally, a phase transition that is not driven by the ion-focus instability is described. There, laser beams are used to mimic a heating process. Here, the radiation pressure force exerted by lasers is exploited to drive equilibrium phase transitions in clusters. By applying random "kicks" from the laser to the dust particles a random change of the velocity of the particles is initiated. This mimics a heating process since the amplitude of the particles' random motion, i.e., their kinetic temperature is effectively increased by these kicks. Such laser-heating of the dust ensemble has been demonstrated for a number of 2D and 3D dust systems, see e.g. [33–36].

To investigate the melting in 3D dust clusters, such a cluster was trapped in a glass box that was placed on top of the electrode of an RF discharge as described in Sect. 8.4. In this particular case, 60 particles of 4.86 µm diameter are confined to form a Yukawa ball [37]. The 3D positions and trajectories were measured using a stereoscopic setup using three cameras, see Fig. 8.15a and Sect. 10.6. An expanded beam of a laser was used to illuminate the particles from two sides. For laser-heating the dust cluster to melting, two laser beams from diode lasers are used that are focused to the Yukawa ball from two opposing sides. The two beams are swept randomly over the cross section of the cluster by galvanometric mirrors. Particles

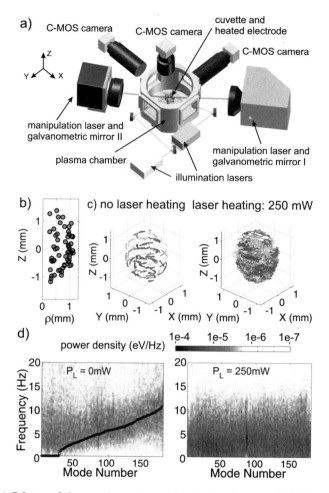

Fig. 8.15 (**a**) Scheme of the experimental setup for the laser-heating of 3D dust clusters. (**b**) Particle positions in cylindrical $\rho - z$ coordinates indicating the two-shell structure of this cluster with $N = 60$ particles. (**c**) Trajectories of the cluster without laser-heating and with laser heating at a laser power of 250 mW. (**d**) Measured mode spectrum of the cluster without laser-heating and with laser heating. From [37]

that are hit by the laser experience a force for a brief moment and are pushed in the direction of the laser beam. Due to the random laser motion the particles are randomly accelerated and, hence, a heating process is mimicked. With increased laser power the particles are heating more strongly. With such a setup Yukawa balls with particle numbers $N < 100$ have been heated into the liquid regime.

The Yukawa ball with $N = 60$ particles is spherical in shape and consists of two shells, see Fig. 8.15b. The 3D trajectories of the observed particles in the cluster were recorded for over 30,000 frames (corresponding to 300 s). The unheated cluster shows a slow rotation of the cluster as a whole, the cluster is found in an ordered state during the total sequence. However, by applying a laser-heating with a laser power of 250 mW for each of the two heating beams, the cluster shows an unordered, liquid-like behavior, see Fig. 8.15c. The melting process can be quantitatively characterized using statistical measures as correlation functions or configurational entropies [36, 38, 39].

Here, we like to look into the normal modes which are shown in Fig. 8.15d. For the unheated cluster, the mode spectrum is that of an ordered system where the regions of highest spectral power density closely follow the expected mode frequencies. For each mode we find a relatively sharp peak in the spectrum, the spectral width is small since the cluster is trapped a relatively low gas pressure, so that the modes are not overdamped.

In contrast, for the heated cluster, the energy in the modes is not located in a narrow frequency band as for the ordered cluster. High spectral power densities are found in a broad frequency band with frequencies up to 10 Hz for all modes. Most of the energy in the system is seen to be located in a range with its maximum between 1 and 5 Hz. Frequent particle exchanges occurring in this liquid regime prevent the establishment of normal mode oscillations. Thus, the spectrum is much broader as for the cluster in the solid phase. Not unexpected, this shows that the mode analysis presented here is mainly suited for the study of ordered systems.

References

1. J.J. Thomson, Philos. Mag. **39**, 237 (1904)
2. A. Mortensen, E. Nielsen, T. Matthey, M. Drewsen, Phys. Rev. Lett. **96**, 103001 (2006)
3. R.W. Hasse, Phys. Rev. Lett. **90**, 204801 (2003)
4. P. Leiderer, W. Ebner, V.B. Shikin, Surf. Sci. **113**, 405 (1982)
5. L.P. Kouwenhoven, D.G. Austing, S. Tarucha, Rep. Prog. Phys. **64**(6), 701 (2001). http://stacks. iop.org/0034-4885/64/i=6/a=201
6. M. Bonitz, C. Henning, D. Block, Rep. Prog. Phys. **73**(6), 066501 (2010). http://stacks.iop.org/ 0034-4885/73/i=6/a=066501
7. G.J. Kalman, J.M. Rommel, K. Blagoev (eds.), *Strongly Coupled Coulomb Systems* (Plenum Press, New York, 1998)
8. A. Melzer, B. Buttenschön, T. Miksch, M. Passvogel, D. Block, O. Arp, A. Piel, Plasma Phys. Controlled Fusion **52**, 124028 (2010)
9. A. Homann, A. Melzer, S. Peters, R. Madani, A. Piel, Phys. Rev. E **56**, 7138 (1997)
10. B. Liu, K. Avinash, J. Goree, Phys. Rev. Lett. **91**, 255003 (2003)
11. A. Melzer, Phys. Rev. E **73**, 056404 (2006)
12. T.E. Sheridan, K.D. Wells, Phys. Rev. E **81**, 016404 (2010)
13. W.T. Juan, Z.H. Huang, J.W. Hsu, Y.J. Lai, L. I, Phys. Rev. E **58**(6), 6947 (1998)
14. M. Klindworth, A. Melzer, A. Piel, V. Schweigert, Phys. Rev. B **61**, 8404 (2000)
15. F.M.H. Cheung, C. Brunner, A.A. Samarian, B.W. James, AIP Conf. Proc. **799**, 185 (2005)
16. A. Melzer, Phys. Rev. E **67**, 016411 (2003)
17. V.M. Bedanov, F. Peeters, Phys. Rev. B **49**, 2667 (1994)
18. O. Arp, D. Block, M. Klindworth, A. Piel, Phys. Plasmas **12**, 122102 (2005)
19. S. Käding, D. Block, A. Melzer, A. Piel, H. Kählert, P. Ludwig, M. Bonitz, Phys. Plasmas **15**, 073710 (2008)
20. O. Arp, D. Block, A. Piel, A. Melzer, Phys. Rev. Lett. **93**, 165004 (2004)

21. T. Antonova, B.M. Annaratone, D.D. Goldbeck, V. Yaroshenko, H.M. Thomas, G.E. Morfill, Phys. Rev. Lett. **96**, 115001 (2006)
22. R.W. Hasse, V.V. Avilov, Phys. Rev. A **44**, 4506 (1991)
23. M. Bonitz, D. Block, O. Arp, V. Golubnychiy, H. Baumgartner, P. Ludwig, A. Piel, A. Filinov, Phys. Rev. Lett. **96**, 075001 (2006)
24. D. Block, S. Käding, A. Melzer, A. Piel, H. Baumgartner, M. Bonitz, Phys. Plasmas **15**, 040701 (2008)
25. P. Ludwig, S. Kosse, M. Bonitz, Phys. Rev. E **71**, 046403 (2005)
26. S. Apolinario, B. Partoens, F. Peters, New J. Phys. **9**, 283 (2007)
27. Y.J. Lai, L. I, Phys. Rev. E **60**(4), 4743 (1999)
28. C. Henning, H. Baumgartner, A. Piel, P. Ludwig, V. Golubnichiy, M. Bonitz, D. Block, Phys. Rev. E **74**, 056403 (2006)
29. D. Block, M. Kroll, O. Arp, A. Piel, S. Käding, Y. Ivanov, A. Melzer, C. Henning, H. Baumgartner, P. Ludwig, M. Bonitz, Plasma Phys. Control. Fusion **49**, B109 (2007)
30. A. Piel, J.A. Goree, Phys. Rev. E **88**, 063103 (2013). https://doi.org/10.1103/PhysRevE.88. 063103.
31. V.A. Schweigert, F. Peeters, Phys. Rev. B **51**, 7700 (1995)
32. R. Ichiki, Y. Ivanov, M. Wolter, Y. Kawai, A. Melzer, Phys. Rev. E **70**, 066404 (2004)
33. M. Wolter, A. Melzer, Phys. Rev. E **71**, 036414 (2005)
34. V. Nosenko, J. Goree, A. Piel, Phys. Plasmas **13**, 032106 (2006)
35. J. Schablinski, D. Block, A. Piel, A. Melzer, H. Thomsen, H. Kählert, M. Bonitz, Phys. Plasmas **19**, 013705 (2012)
36. A. Schella, T. Miksch, A. Melzer, J. Schablinski, D. Block, A. Piel, H. Thomsen, P. Ludwig, M. Bonitz, Phys. Rev. E **84**, 056402 (2011)
37. A. Schella, M. Mulsow, A. Melzer, H. Kählert, D. Block, P. Ludwig, M. Bonitz, New J. Phys. **15**, 113021 (2013)
38. A. Schella, M. Mulsow, A. Melzer, J. Schablinski, D. Block, Phys. Rev. E **87**, 063102 (2013)
39. M. Mulsow, A. Melzer, Phys. Rev. E **96**, 053202 (2017). https://doi.org/110.1103/PhysRevE. 96.053202

Chapter 9
Dusty Plasmas and Magnetic Fields

The investigation of dusty plasmas under the influence of magnetic fields has attracted great interest over the recent years. The main driver of these investigations is, of course, to study the magnetization effects on the kinetic level of individual particles and to make these effects visible using the dust particles as the magnetized species.

So far, the goal of identifying magnetization of the dust species on the kinetic level has not been clearly demonstrated. A comprehensive and well-founded description of the various aspects of dusty plasmas under magnetic fields is difficult since experiments are still rare. The interpretation of experimental results is demanding due to the complex role of magnetized electrons and ions. Nevertheless, in the following, a few aspects of experimental situations of dusty plasmas under magnetic fields will be presented and discussed. For a more elaborate overview, see e.g. [1, 2].

9.1 Magnetization Parameters

When a magnetic field $\mathbf{B}$ is present, the charged plasma species (electron, ion and dust) are subject to the Lorentz force in addition to the other forces $\mathbf{F}$ described in Chap. 3. Then, the equation of motion can be written as

$$m_\alpha \ddot{\mathbf{r}}_\alpha = Q_\alpha \mathbf{v}_\alpha \times \mathbf{B} + \mathbf{F}. \tag{9.1}$$

Here, $\alpha = $ e, i, d denotes the species—electrons, ions, dust—with charge Q_α and mass m_α. In the absence of other forces, the particles perform gyro-orbits with the so-called Larmor radius

$$r_\alpha = \frac{m_\alpha v_\alpha}{Q_\alpha B}.$$

© Springer Nature Switzerland AG 2019
A. Melzer, *Physics of Dusty Plasmas*, Lecture Notes in Physics 962,
https://doi.org/10.1007/978-3-030-20260-6_9

The relevance of the Lorentz force, and hence the magnetization of a plasma component (electron, ion or dust), can be described by different parameters. A first parameter is the so-called Hall parameter, derived from the comparison of friction force and Lorentz force,

$$h_\alpha = \frac{\omega_{c\alpha}}{\nu_{\alpha n}}. \tag{9.2}$$

It relates the cyclotron frequency

$$\omega_{c\alpha} = \frac{Q_\alpha B}{m_\alpha} \tag{9.3}$$

of the species α to its collision frequency $\nu_{\alpha n}$ with the neutral gas background of the plasma. Magnetic field effects can be expected when h_α becomes of the order of 1, i.e. when the particle can perform (a significant fraction of) a gyro-orbit before being perturbed by a collision.

For dust particles, where the charge is proportional to the particle radius a according to the OML model, see Eq. (2.21), and the collision frequency is given by the Epstein drag frequency $\nu_{dn} = \beta$ according to Eq. (3.15), the Hall parameter scales as (see also [1])

$$h_\alpha \sim \frac{a\,B}{a^3}\frac{a}{p} = \frac{B}{a\,p},$$

where, p is the gas pressure in the discharge.

A second parameter, the magnetization, is derived from comparing the electrostatic time scale given by the plasma frequency of species α and its gyro-frequency (cyclotron frequency). It can, hence, be defined as the ratio of these two frequencies as

$$M_\alpha = \frac{\omega_{c\alpha}}{\omega_{p\alpha}}. \tag{9.4}$$

This parameter is particularly suited for the characterization of the collective modes in magnetized plasmas [3]. With the plasma frequency defined in Eq. (6.1) and Eq. (6.2), it is seen that

$$M_\alpha \sim \frac{B}{\sqrt{m_\alpha}}.$$

Form these quantities it follows that the light electrons will first be magnetized with $h_e > 1$ and $M_e > 1$ at magnetic fields B as low as a few Milliteslas. For the heavier ions to be magnetized ($h_i > 1$ and $M_i > 1$), usually much higher magnetic fields of the order of 100 mT and larger are required. Such high fields have been achieved in dusty plasma experiments using water-cooled [4] or superconducting magnets [5–8] leading to a number of interesting results that are indicated below.

Finally, for the dust particles to be magnetized, the quantity B/a has to be maximized to achieve $h_d > 1$ and $M_d > 1$. Presently available homogeneous and steady magnetic fields range up to about 5 T. Yet, the particle size is limited to a minimum of about 50 to 100 nm, since we still want to detect and observe the dust particles, see Chap. 10. It is estimated that such small dust particles could just be magnetized [1] at the highest fields. At these field strengths inevitably also the electrons and ions are magnetized.

Hence, from the point of view of the dust dynamics, a real magnetized dusty plasma is achieved only in this high-field stage. However, already the magnetization of the electrons and, at higher fields, the ions, leads to decisive changes of the discharge properties which will then in turn influence the dust dynamics. These different aspects will be discussed in the following.

9.2 Moderate Axial Fields: Rotation

We start with the discussion with dusty plasmas at low to moderate magnetic fields where the magnetic field is applied along the discharge axis and along the gravitational acceleration. The experiments have generally been performed with micrometer-sized dust particles. At these moderate fields, magnetization effects on the dust are expected to be small i.e. $h_d \ll 1$ and $M_d \ll 1$. However, it is observed that ensembles of trapped dust particles begin to rotate as a whole around the magnetic field axis, see Fig. 9.1 or [5, 9–11]. The rotation direction usually is in

Fig. 9.1 A cloud of dust particles rotating under the influence of an axial magnetic field. Reprinted from [5], with the permission of AIP Publishing

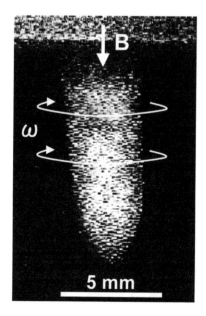

the $\mathbf{E} \times \mathbf{B}$ direction where $\mathbf{E}$ is the (horizontal, radial) electric field to confine the dust particles and $\mathbf{B}$ is the vertical magnetic field. In the particular experiment of Fig. 9.1, a dust cloud trapped below the floating grid in a DC plasma discharge is found to rotate at a frequency Ω of the order of $0.1\,\mathrm{s}^{-1}$ at a field of 40 mT [5]. A similar behavior has been observed for particles trapped in the sheath of RF discharges [10, 11].

The origin of the rotation lies in the $\mathbf{E} \times \mathbf{B}$ motion of the ions that is eventually transferred to the dust by the ion drag force. For a more quantitative analysis, we consider a situation where, in the vertical direction, the dust particles are simply confined by the electric field force due to the vertical sheath electric field E_z and gravity, i.e. $Q_d E_z = m_d g$. Radially, the dust is confined by a radial electric field E_r, e.g. due to a barrier on the electrode (compare e.g. Sect. 8.3). Then, the ion motion in a plasma where ion-neutral collision are frequent is given by

$$\mathbf{u}_i = \mu_i(\mathbf{E} + \mathbf{u}_i \times \mathbf{B}), \tag{9.5}$$

where μ_i is the ion mobility. Under the action of a vertical magnetic field B_z, the radial ion velocity component can be written as $u_{ir} = \mu_i E_r$ and the azimuthal component as [10]

$$u_{i\phi} = \mu_i^2 E_r B_z. \tag{9.6}$$

With this azimuthal ion velocity $u_{i\phi}$ in $E_r \times B_z$ direction, the dust is driven via the ion drag force, Eq. (3.10).

In experiments at very low gas pressure [9] this direct mechanism is found quantitatively compatible with the observed rotation. However, for typical plasma experiments with gas pressures around a few Pascal such a direct ion-drag driven interaction usually seems to be too weak to produce the observed rotation speeds [11]. In these situations, one finds a quantitative agreement when the azimuthally drifting ions are assumed to excite, in the first place, a rotation of the entire neutral gas background that in turn drives the dust particle rotation via the neutral gas drag, Eq. (3.14). Hence, neutral gas motion plays an important role in the rotation of dust clusters in a magnetic field [11]

In a recent experiment [12] with magnetic fields of the order of 1 T the rotation of a two-dimensional dust system is exploited to study phase transitions in these dust clouds. At low magnetic field, the dust cloud was found to rotate as a rigid body, where an ordered state was maintained. At higher fields the shear in the drive leads to a shear motion within the dust cloud that eventually disrupts the ordered dust arrangement.

9.3   Moderate Horizontal Fields: Sideways Motion

In the following, a situation is discussed where a magnetic field B_x is applied horizontally, i.e. parallel to the electrodes, see Fig. 9.2a. There, one finds a quite complex reaction of the dust particles under small fields where only electrons are magnetized and still direct magnetic field effects on the dust can be neglected [13].

Again, the dust is trapped by the force balance of vertical electric field force and gravity $Q_d E_z = m_d g$. Now the reaction of the plasma components (mainly electrons) to the $E_z \times B_x$ drift with the vertically confining electric field E_z and the horizontal magnetic field B_x becomes important.

In an experiment, upon increasing the magnetic field, the dust particles in the sheath are found to first move sideways against the $\mathbf{E}_z \times \mathbf{B}_x$ direction. At further increased magnetic field the particles start to move back along the $\mathbf{E}_z \times \mathbf{B}_x$ direction, see Fig. 9.2b. The details of this sideways motion of the dust depend quite strongly on the particle size as well as the plasma parameters, gas pressure etc. [13].

This behavior can be understood by looking at the dynamics of the electrons and ions in the sheath, see e.g. [14, 15]. First, at the gas pressures used in the experiment the electrons are magnetized ($h_e > 1$) already at fields as low as $B = 1\,\text{mT}$, the ions can be considered unmagnetized with $h_i < 1$. The magnetic field induces an $\mathbf{E} \times \mathbf{B}$-drift of the magnetized electrons. Since the ions are left behind, an ambipolar electric field in $\mathbf{E} \times \mathbf{B}$ direction is evoked, see Fig. 9.2a. This ambipolar field leads to an electric field force opposite to $\mathbf{E} \times \mathbf{B}$ direction for the negatively charged dust. In contrast, the ions are forced to move in $\mathbf{E} \times \mathbf{B}$ direction and lead to an ion drag force in that direction which, as discussed above, is enhanced by neutral gas drag driven by the ion-neutral collisions. The complicated force balance between these forces determines the motion of the dust particles. A quantitative analysis using models of the magnetized sheath [15] and the different forces (electric field force, ion and neutral drag) yields good agreement with the observations [16].

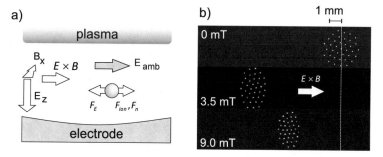

Fig. 9.2 (a) Sketch of the setup with a horizontal magnetic field. (b) Top view of a dust cluster of $7\,\mu\text{m}$ particles under different magnetic field strengths at a gas pressure of 5 Pa. From [3]

9.4   Strong Axial Fields and Microparticles

Experiments on dusty plasmas at very high magnetic field strengths ($B > 1\,\mathrm{T}$) are rare and not very well understood, so far. In these experiments usually a strong axial (vertical) magnetic field is applied and the RF discharge electrodes are perpendicular to the field. The ions become magnetized at these fields ($h_i > 1$), the dust Hall parameter can be estimated to be of the order of $h_d \approx 10^{-4}$, so that direct dust magnetization still is not expected.

By magnetization of the electrons and ions, the magnetic field introduces a directionality to the plasma. This directionality manifests in different ways that will be illustrated in the following.

Plasma Filamentation Under strong magnetic fields the plasma is found to undergo an instability where a filamentation of the plasma occurs, see e.g. [6] and Fig. 9.3. These filaments are regions of elevated plasma density which are quite localized in the horizontal direction but extended along the magnetic field lines. Their formation seems to be independent of the presence of dust. The filaments occur in different types of gases, depend somewhat on the plasma power and seem to favor low gas pressures. The filaments heavily influence the dust motion, the dust can even be used to trace the outlines of the filaments [6]. In dense dust clouds the filamentation, however, seems to be suppressed [4].

Mapping As another effect of the directionality introduced by the magnetic field, it is found that structures of the electrode can be mapped onto dust structures [8]. As an example, an upper electrode consisting of a square wire mesh imposes its structure onto the dust particle cloud trapped in the sheath of the lower electrode some centimeters below the upper wire mesh. Then, the dust particles also arrange in a square pattern with a distance prescribed by the mesh. The formation of this pattern occurs under sufficient magnetic field (of the order of 1 T) and is favored at lower gas pressures. Also it is seen that the dust particles preferably move along the imposed wire mesh structure [8].

Fig. 9.3 Filamentation of an argon RF discharge at a magnetic field of $B = 1.6\,\mathrm{T}$. The filaments are seen as vertical stripes with increased light emission. Reprinted figure with permission from [6]. © (2011) by the American Physical Society

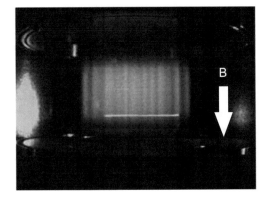

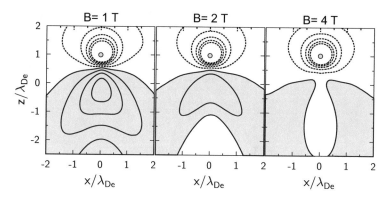

Fig. 9.4 Simulation of the collisionless wake of a dust particle in an argon plasma for different magnetic field strengths: $B = 1$ T, 2 T and 4 T, respectively. The potential contours are indicated (solid lines: positive potentials, dashed lines: negative potentials). The regions of positive (attractive) potential have been highlighted. Note the different vertical scale with respect to Fig. 4.2. After [23], figure by courtesy of A. Piel

Wakefield As introduced in Sect. 4.2, the ions streaming past a dust particle in the plasma sheath form a wakefield structure downstream of the dust particle. In the unmagnetized case, the ions are scattered into the downstream region by the dust electric field where a region of enhanced positive space charge is formed: the ion focus or wakefield [17–20].

The wakefield under strong magnetic fields has been studied in recent simulations [21–24]. In particle-in-cell simulations one can follow and study the motion of ions under magnetic fields taking the dust-charging into account self-consistently [22], and even physically realistic dimensions and ion-to-electron mass ratios [23, 24]. These simulations reveal that, in the collisionless case, under modest magnetic fields (up to e.g. $B = 1$ T) the ion wakefield is somewhat weakened, see also [25], and the wake structures move closer to the dust particle, see Fig. 9.4 and compare Fig. 4.2. With larger fields ($B = 2$ T), regions of reduced ion density appears downstream of the dust particle ("ion shadow"). This is since, at these magnetic field strengths, the ions are magnetized and follow the magnetic field lines. Hence, they cannot be effectively scattered into the region downstream of the dust. Consequently, that region becomes depleted of ions forming now an ion shadow instead of an ion focus. This ion shadow is even more pronounced at stronger fields $B = 4$ T. In the same way as the ion shadow appears and increases the attractive part of the wake is pushed outwards. Consequently, the non-reciprocal horizontal attraction on a second, lower-layer particle is maintained at low field strengths, but reverses to a repulsive force at higher field strengths [23]. Under the presence of ion-neutral collisions the transition from an attractive to a repulsive wake (ion focus to ion shadow) occurs in an argon plasma only at decisively higher magnetic fields of about 10 T [24].

Experiments on the magnetized wakefield interaction of microparticles are tedious and rare. In the conducted experiments [26], the wakefield interaction has been probed by a pair of vertically aligned particles and the mutual interaction between them is derived from the resonance technique. These experiments show that the interaction between the two dust particles becomes weaker with increasing field supporting the weakening of the wakefield strength. However, the coupling from the lower to the upper particle remains always stronger than vice versa up to the maximum applied field of 2.5 T, thus favoring the aligned situation up to those fields.

9.5   Strong Axial Fields and Nanoparticles

As has been described above, the way towards a plasma containing magnetized dust, i.e. a plasma with $h_d > 1$ and/or $M_d > 1$, requires the use of small particles, the smaller the better [1]. However, optical diagnostics of dust systems can be applied only for particles with diameters larger than about 50 nm.

So far, very few experiments [4] exist where submicron dust is studied under magnetic fields and truly magnetized dust systems have not been reported, so far. One of the difficulties of studying such systems is that the plasma properties and the situation of dust confinement drastically changes with the presence of a magnetic field. Filamentation of the plasma as discussed above is one example.

In an approach to generate magnetized dusty plasmas with submicron particles [4], dust particles have been grown in argon-acetylene plasmas without magnetic field (compare Sect. 11.1). Then, after the particles have reached a certain size of about 150 nm in radius, dust growth is stopped by switching off the acetylene flow. Afterwards, a magnetic field of up to 500 mT is applied to the dusty plasma. It is found that the dust cloud becomes more restricted and confined to the space between the electrodes, see also Fig. 9.5. This is on the one hand directly due to a change of the plasma properties where due to the magnetization of electrons (and partly ions) the plasma is confined to the region between the electrodes. On the other hand, a rotary dust motion excited by the Hall component of the ion drag force (see Sect. 9.2) changes the force balance for the dust particles [4].

Fig. 9.5 Dust cloud of plasma-grown particles of about 150 nm radius in the unmagnetized (left) and the magnetized (right) plasma (inverted gray scale). Reprinted from [4], with the permission of AIP Publishing

9.6 Roundup

Already small and moderate magnetic fields introduce indirect effects on the dust system: the magnetized electrons (and sometimes ions) perform $\mathbf{E} \times \mathbf{B}$ drifts or gyro-motions. Thereby, they influence mainly electric field forces or ion drag that then drive a motion of the charged dust particles. In many situations the ion motion, due to collisions with the neutral gas, also sets up a motion of the gas background that amplifies the ion drag force via the neutral drag.

At stronger fields with magnetized electrons and ions the plasma structures become localized perpendicular to the field due to the small gyro-orbits of the electrons and ions. The structures are elongated along the field leading to the filamentation of the plasma and the imposed grid structures and, on the microscopic level, to the modified wake structure with the appearance of and ion shadow.

Dust systems where electrons, ions and dust are magnetized require very strong fields and submicron dust particles. So far, clear evidences for magnetized dust systems are lacking.

Aspects that have not been discussed here include, e.g., dust tori in anodic plasmas [27–29], experiments with paramagnetic dust particles [30–32], or self-excited oscillations due to delayed charging at strong fields [33]. A peculiar way to achieve "magnetization" of the dust component is by forcing the dust system to rotate without applying any magnetic field. There, one exploits the equivalence of the Lorentz force on the dust and the Coriolis force on a rotating dust system [34, 35]. This pseudo-magnetization of the dust is an artificial situation where the electrons and ions are left unmagnetized. The above mentioned situations are described in some more detail in, e.g., [2].

References

1. E. Thomas, R.L. Merlino, M. Rosenberg, Plasma Phys. Controlled Fusion **54**, 124034 (2012)
2. D. Block, A. Melzer, J. J. Phys. B: At., Mol. Opt. Phys. **52**, 063001 (2019). https://doi.org/10.1088/1361-6455/ab023f
3. H.Kählert, A. Melzer, M. Puttscher, T. Ott, M. Bonitz, Eur. Phys. J. D **72**, 83 (2018). https://doi.org/10.1140/epjd/e2017-80409-x
4. B. Tadsen, F. Greiner, A. Piel, Phys. Plasmas (1994-present) **21**, 103704 (2014)
5. N. Sato, G. Uchida, T. Kaneko, S. Shimizu, S. Iizuka, Phys. Plasmas **8**, 1786 (2001)
6. M. Schwabe, U. Konopka, P. Bandyopadhyay, G.E. Morfill, Phys. Rev. Lett. **106**, 215004 (2011). https://doi.org/10.1103/PhysRevLett.106.215004
7. J. Carstensen, F. Greiner, D. Block, J. Schablinski, W.J. Miloch, A. Piel, Phys. Plasmas **19**, 033702 (2012)
8. E. Thomas, B. Lynch, U. Konopka, R.L. Merlino, M. Rosenberg, Phys. Plasmas (1994-present) **22**, 030701 (2015). http://dx.doi.org/10.1063/1.4914089.
9. S. Nunomura, N. Ohno, S. Takamura, Jpn. J. Appl. Phys. **36**, 877 (1997)
10. U. Konopka, D. Samsonov, A.V. Ivlev, J. Goree, V. Steinberg, G. Morfill, Phys. Rev. E **61**, 1890 (2000)
11. J. Carstensen, F. Greiner, L.J. Hou, H. Maurer, A. Piel, Phys. Plasmas **16**, 013702 (2009)

12. S. Jaiswal, T. Hall, S. LeBlanc, R. Mukherjee, E. Thomas, Phys. Plasmas **24**(11), 113703 (2017). https://doi.org/10.1063/1.5003972
13. M. Puttscher, A. Melzer, Phys. Plasmas (1994-present) **21**, 123704 (2014)
14. Y. Maemura, S.C. Yang, H. Fujiyama, Surf. Coat. Technol. **98**, 1351 (1998)
15. B.P. Pandey, S.V. Vladimirov, A. Samarian, Phys. Plasmas **18**(5), 053703 (2011). https://doi.org/10.1063/1.3589799
16. A. Melzer, M. Puttscher, Phys. Plasmas **24**(5), 053701 (2017). https://doi.org/10.1063/1.4981928
17. S.V. Vladimirov, M. Nambu, Phys. Rev. E **52**, R2172 (1995)
18. M. Nambu, S.V. Vladimirov, P.K. Shukla, Phys. Lett. A **203**, 40 (1995)
19. V.A. Schweigert, I.V. Schweigert, A. Melzer, A. Homann, A. Piel, Phys. Rev. E **54**, 4155 (1996)
20. A. Melzer, V.A. Schweigert, I.V. Schweigert, A. Homann, S. Peters, A. Piel, Phys. Rev. E **54**, R46 (1996)
21. J.P. Joost, P. Ludwig, H. Kählert, C. Arran, M. Bonitz, Plasma Phys. Controlled Fusion **57**, 025004 (2015). http://stacks.iop.org/0741-3335/57/i=2/a=025004
22. W.J. Miloch, D. Darian, M. Mortensen, Phys. Scr. **92**, 114006 (2017). http://stacks.iop.org/1402-4896/92/i=11/a=114006
23. A. Piel, F. Greiner, H. Jung, W.J. Miloch, Phys. Plasmas **25**, 083702 (2018). https://doi.org/10.1063/1.5039587
24. A. Piel, H. Jung, F. Greiner, Phys. Plasmas **25**, 083703 (2018). https://doi.org/10.1063/1.5039606
25. M. Nambu, M. Salimullah, R. Bingham, Phys. Rev. E **63**, 056403 (2001). https://doi.org/10.1103/PhysRevE.63.056403
26. J. Carstensen, F. Greiner, A. Piel, Phys. Rev. Lett. **109**, 135001 (2012)
27. I. Pilch, T. Reichstein, A. Piel, Phys. Plasmas **15**(10), 103706 (2008). https://doi.org/10.1063/1.3006085
28. T. Reichstein, J. Wilms, F. Greiner, A. Piel, A. Melzer, Contrib. Plasma Phys. **52**, 813 (2012)
29. A. Piel, D. Block, A. Melzer, M. Mulsow, J. Schablinski, A. Schella, F. Wieben, J. Wilms, Eur. Phys. J. D **72**(5), 80 (2018). https://doi.org/10.1140/epjd/e2017-80371-7.
30. D. Samsonov, S. Zhdanov, G. Morfill, V. Steinberg, New J. Phys. **5**, 24 (2003)
31. V.V. Yaroshenko, G.E. Morfill, D. Samsonov, Phys. Rev. E **69**, 016410 (2004)
32. M. Puttscher, A. Melzer, New J. Phys. **16**, 043026 (2014)
33. M. Puttscher, A. Melzer, U. Konopka, S. LeBlanc, B. Lynch, E. Thomas Jr., Phys. Plasmas **24**(1), 013701 (2017). https://doi.org/10.1063/1.4973231
34. H.Kählert, J. Carstensen, M. Bonitz, H. Löwen, F. Greiner, A. Piel, Phys. Rev. Lett. **109**, 155003 (2012)
35. P. Hartmann, Z. Donkó, T. Ott, H. Kählert, M. Bonitz, Phys. Rev. Lett. **111**, 155002 (2013). https://doi.org/10.1103/PhysRevLett.111.155002

Chapter 10
Diagnostic Methods in Dusty Plasmas

As we have seen, laboratory experiments on dusty plasmas usually make use of solid particles in the size range between nanometers to micrometers in diameter immersed in a gaseous plasma environment. The dust size, the interparticle distance and the time scales associated with the particle motion are ideally suited to study the dust by optical techniques.

Here, we like to review different diagnostic methods to measure the structure and dynamics of two-dimensional and three-dimensional micro-dust systems. Also, approaches to determine the dust size (distribution) and dust density (distribution) in a nano-dust cloud are presented.

10.1 Light Scattering from Particles

The easiest way to image dust particles is to illuminate the particles with a thin sheet of laser light and record the scattered light with a video camera, see e.g. Fig. 10.1. Typically, a laser sheet with a thickness of less than one millimeter is produced using a cylindrical lens (system) in front of a laser. The light scattered by the illuminated particles is often recorded by video cameras under right angle. The well defined wavelength λ of the laser allows to use interference filters in front of a camera to block most of the plasma glow allowing only the scattered light at wavelength λ to be imaged by the camera. Such a system is referred to as a video microscope.

At first glance this setup seems simple and flexible, but one has to consider some important aspects. The scattering of light from a particle is a relatively complex problem and is described by Mie scattering theory [1]. Figure 10.2a shows the angular intensity distribution of light scattered from a particle of $a = 1.5\,\mu m$ radius illuminated by a laser at a wavelength of $\lambda = 532\,nm$. Here, also the different polarizations of the laser with respect to the scattering plane are considered (we will return to that below in Sect. 10.7.1). In both polarization states, the scattering

© Springer Nature Switzerland AG 2019
A. Melzer, *Physics of Dusty Plasmas*, Lecture Notes in Physics 962,
https://doi.org/10.1007/978-3-030-20260-6_10

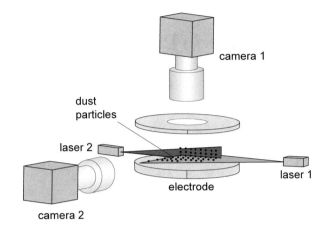

Fig. 10.1 Schematic setup for a simple video microscope. The dust particles are illuminated by a horizontal and/or a vertical laser sheet. The scattered laser light is recorded at right angle

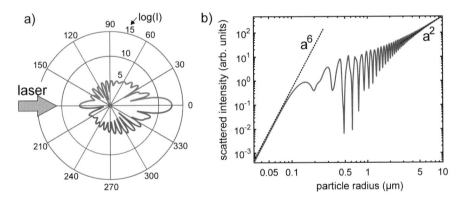

Fig. 10.2 (**a**) Angle-resolved scattering intensity of a particle of 1.5 μm radius illuminated by a laser at a wavelength of 532 nm. The red (blue) line corresponds to scattering with perpendicular (parallel) polarization. The scattered intensity is on a logarithmic scale. (**b**) (Perpendicular) scattering intensity of laser light at 532 nm under an angle of 90° as a function of particle radius

intensity shows a strong variation with the scattering angle. Most of the light is scattered in forward direction, the light scattered under 90° has a relatively small intensity.

Moreover, for a fixed observation angle, the scattered intensity drastically depends on the particle radius as seen in Fig. 10.2b. For a particle much smaller than the illumination wavelength, $a \ll \lambda$ (Rayleigh regime), the scattering intensity scales with a^6. This very strong size dependence is the reason why particles smaller than about 50 nm cannot be analyzed by light scattering techniques. In contrast, for particles much larger than the wavelength ($a \gg \lambda$), the scattering intensity scales just with the geometric cross section πa^2. In the (interesting) intermediate size range, the scattering intensity oscillates strongly with particle size.

CCD cameras and modern CMOS[1] cameras are sensitive enough to record the scattered light of micron-sized particles even with small exposure times of the order of a few milliseconds. With that, frame rates of a few hundred frames (images) per second (fps) can easily be realized even at megapixel resolution.[2] The camera lens is an important factor for recording high-quality images. Usually, the camera is placed outside the experimental chamber and is thus some centimeters away from the dust system. Moreover, typically, the camera sensor and the desired field of view are about the same size. Hence, the magnification factor is close to one and thus macro lenses have to be used since they can be focused even at small working distance. One should use lenses with large maximum aperture (small minimum f-number such as f/1.4 or f/2.8), so-called "fast lenses", because they deliver higher light intensity and thus allow shorter (faster) exposure times. The opening aperture determines the depth of field which is an important factor in the diagnostics of 3D systems, as will be discussed below.

10.2 Image Analysis

Once images have been recorded, reliable image processing and analysis is required.[3] A typical image of a dust cloud has a dark background with a few brighter clusters of pixels representing the particles. In general, the number of pixels representing a particle and separation between them depends on the camera resolution, the magnification of the macro lens, the degree of focussing and the brightness of the particles, i.e. it strongly depends on the experimental setup and equipment. Generally, due to its scattering properties, a particle appears larger on the camera than its actual size and the particle image for microparticles covers a few pixels.

To extract precise particle positions from these images several algorithms have been proposed that allow the determination of positions even with subpixel resolution. However, when evaluating data [2, 3], distortions of measured velocity distributions were found that originate from an effect called pixel locking. Pixel locking is an artifact where calculated particle positions are preferably found at certain locations relative to pixel edges. Thus, especially for high precision experiments in dusty plasmas like normal mode analysis, dust kinetic temperature

---

[1]CCD: charge-coupled device, CMOS: complementary metal-oxide-semiconductors. In former years, CCD cameras have been used in situations where high-quality image data are needed. However, over the recent years, CMOS cameras, that provide higher data rates, have considerably improved in image quality and nowadays generally outperform CCD cameras.

[2]At present, standard USB 3.0 allows a data transfer rate of about 300 MByte/s allowing 300 fps at 1 Megapixel resolution, but also faster techniques are available.

[3]The recorded video images should be stored in a lossless data format. The use of lossy formats, such as "jpeg", is strongly discouraged since they introduce artifacts that interfere with image analysis.

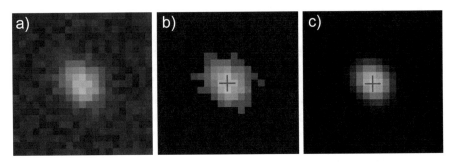

Fig. 10.3 Illustration of the working principle of the particle positioning methods. (**a**) Artificial particle image with noise on the camera sensor. (**b**) Particle image according to moment method. All pixels above a threshold value are considered to belong to the particle, pixels below threshold are ignored. (**c**) Artificial image after treatment with a Gaussian bandpass filter. The crosses indicate the determined particle position

measurements or wave dispersion experiments etc., pixel locking can distort the result.

The following paragraphs describe two often used detection methods. Their working principle is illustrated in Fig. 10.3. A detailed description and error analysis with respect to pixel locking of these methods have been given by Ivanov et al. [4] and Feng et al. [5].

Moment Method This method is one of the most widely applied techniques in dusty plasmas. First, a threshold value is chosen for the entire image and all contiguous pixels with intensities above the threshold are considered as a particle, compare Fig. 10.3 a and b. In a second step the "center of mass" is computed for each of these pixel clusters using the pixel intensity to calculate the weighted average. Hence, the mean particle position (X_C, Y_C) in the image is determined from

$$X_C = \frac{\sum_k X_k (I_k - I_{th})}{\sum_k I_k - I_{th}} \quad \text{and} \quad Y_C = \frac{\sum_k Y_k (I_k - I_{th})}{\sum_k I_k - I_{th}}, \qquad (10.1)$$

where I_k is the intensity of pixel k and X_k, Y_k is its x, y-position in the camera image. Here, the threshold intensity is denoted by I_{th} and the sum runs over all contiguous pixels which exceed the threshold. The accuracy of this method has been studied in detail by Feng et al. [5]. They have shown that subpixel resolution is achieved when the recording parameters are well adjusted and the threshold is chosen carefully. Additionally, a detailed guide on how to find the optimum setting for this method and how to avoid pixel-locking is given.

Moment Method with Gaussian Bandpass Filter This algorithm by Crocker et al. originates from the field of colloidal suspensions [6, 7]. In a first step a spatial bandpass filter for smoothing and background subtraction is applied. The width of the Gaussian filter is chosen to match the expected particle size in the image. The result is shown in Fig. 10.3c. After this, contiguous pixels are identified and their weighted center of mass is computed similar to the moment method. The intensity-weighted center (without using a threshold) is calculated over a definite area which is of the size of the particle image. This method allows to achieve subpixel resolution and minimizes pixel locking effects [4]. In addition, the filtering process has the advantage to partially compensate imperfections during the recording process.

Particle Tracking Particle tracking is a process to link particle positions in successive video frames to generate trajectories, i.e. time series of positions $\mathbf{r}_i(t)$ of (each) particle i. The simplest algorithm is the nearest-neighbor approach where the particle j in the actual frame that is closest to the particle i in the previous frame is chosen as the successor of particle i. Refined algorithms use Wiener filters [8] that extrapolates the particle position from the position, velocity and acceleration in the previous frames. Then, the nearest neighbor to this predicted position is searched.

This can be further refined using a Kalman filter approach that in addition to the prediction step also has a correction step. As in the Wiener filter, in the first step previous states are used to predict the current new position. The second step uses the current position measurement to correct the state. For this, a model for the particle motion, such as motion with constant velocity or constant acceleration, is assumed. This Kalman approach is favored when large-scale motions (flows) in the dust cloud are to be resolved.

Velocities and Accelerations From these detection methods the particle positions $\mathbf{r}_i$ of (each) particle i are obtained for consecutive time steps Δt, thus revealing the structure of a given particle arrangement. In many situations also the velocities $\dot{\mathbf{r}}_i = \mathbf{v}_i = \Delta\mathbf{r}_i/\Delta t$ and accelerations $\ddot{\mathbf{r}}_i = \mathbf{a}_i = \Delta\mathbf{v}_i/\Delta t$ are needed to determine the particle dynamics and relevant forces. Here, $\Delta\mathbf{r}_i$ ($\Delta\mathbf{v}_i$) is the difference of particle position (velocity) between two successive video frames, and Δt is the time between these frames, i.e. Δt is the inverse frame rate.

However, despite the possible sub-pixel resolution, the limited accuracy in the determination of the particle positions usually does not allow the direct determination of the velocity by $\mathbf{v}_i = \Delta\mathbf{r}_i/\Delta t$ using the difference between the particle position $\Delta\mathbf{r}_i$ of particle i in successive frames, especially when dealing with small velocities (and, hence, small position differences $\Delta\mathbf{r}_i$). The so determined time series of the velocity would become quite "noisy". Therefore, it is useful first to smooth the original position data by a Savitzky-Golay filter [9] (in many cases a second degree polynomial with a window width of about nine frames seems reasonable, see e.g. [10]). From the smoothed position data, the difference quotient $\mathbf{v}_i = \Delta\mathbf{r}_i/\Delta t$ is computed to obtain the particle velocities.

Similarly, to measure the particle acceleration and thus the forces, the velocity data usually have to be Savitzky-Golay-smoothed again to be used for computing the second difference quotient $\mathbf{a}_i = \Delta\mathbf{v}_i/\Delta t$.

10.3 Particle Image Velocimetry

Particle Image Velocimetry (PIV) is a technique to derive the flow field of an ensemble of (dust) particles. There, instead of deriving the velocity of each particle individually, the average velocity of a group of particles in subfields of the image is determined. This technique is widely applied in fluid dynamics where tracer particles are used that follow the fluid motion [11].

For that purpose, two images with a time difference Δt are considered, e.g. two frames from a video sequence. The images are then divided into several "windows" or "interrogation areas" that contain several to several ten dust particles, see e.g. Fig. 10.4a. Now for each of these interrogation areas, a displacement (velocity) vector is calculated.

This is done by calculating, within each interrogation area, the cross correlation of the area at the two instants, i.e.

$$C(\Delta x, \Delta y) = \sum_{i=-K}^{K} \sum_{j=-L}^{L} I(i, j) I'(i + \Delta x, j + \Delta y). \tag{10.2}$$

Here, I and I' are the pixel intensity values in the interrogation area (of size $K \times L$) at the two instants t and $t + \Delta t$. Then, the two areas are shifted against each other by $(\Delta x, \Delta y)$ and the maximum of the cross correlation C indicates for which $(\Delta x, \Delta y)$ the two images are most similar, see Fig. 10.4b. This determines, how on average the particle ensemble in the interrogation area has shifted from one instant to the next, yielding the velocity vector $\mathbf{v} = (\Delta x / \Delta t, \Delta y / \Delta t)$. From the velocity vector for each area the flow field for the entire image is obtained as in Fig. 10.4c. Further details can be found, e.g. in Ref. [11].

PIV has been introduced for dusty plasmas in the study of the dynamics of dust clouds trapped in an anodic plasma [13, 14]. This technique has been further refined to stereoscopic PIV [15, 16] and tomographic PIV [17].

10.4 Scanning Video Microscopy

The above mentioned video microscopes are very well suited for the diagnostics of particles in a two-dimensional sheet of laser light. However, for the reconstruction of the three-dimensional positions and velocities other approaches have to be followed. Here, a selection of techniques are presented for 3D diagnostics, namely scanning video microscopy [18–20], digital holography [21] and stereoscopic

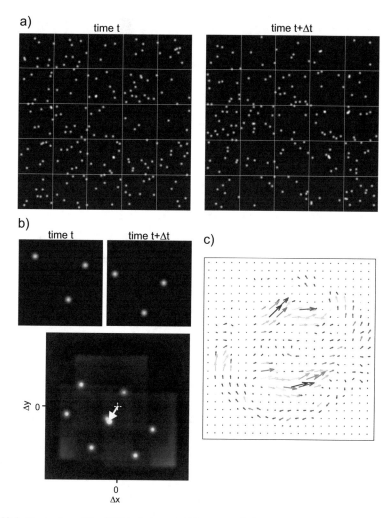

Fig. 10.4 Illustration of the PIV technique. (**a**) Two example images of a dust cloud taken at time t and time $t + \Delta t$. Here, the images are divided into 5×5 interrogation areas. (**b**) Exemplary interrogation area at these two instants together with the their (color-coded) cross correlation. (**c**) Resulting flow field of a dust particle cloud. From [12]

imaging techniques [15, 17, 22–24]. Other techniques include color-gradients in the illumination [25] or plenoptic cameras [26].[4]

---

[4]Plenoptic (or light-field) cameras distribute the light over some pixels so that the direction of the light rays can be recovered. This allows, as post-processing after recording of the light-field image, to calculate images at different focal planes and thus to retrieve 3D information.

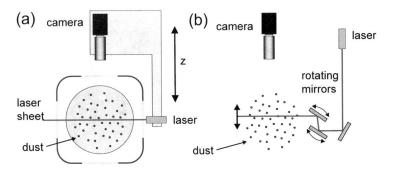

Fig. 10.5 (**a**) Schematic setup of a scanning video microscope. A video microscope is mounted on a translational stage. A scan along the z-axis allows recording of a (stationary) 3D dust cloud. (**b**) Fast scanning video microscope with rotating mirrors. After [20, 27, 28]

Scanning video microscopy is a straightforward extension of the video microscopy discussed above. There, a standard video microscope is mounted on a translation stage together with the illumination laser. The particles are illuminated by a thin laser sheet and the scattered light is detected with a camera at right angle, Fig. 10.5a, just as for the 2D systems. By moving the translation stage a sequence of cross-sections of the trapped dust cloud is recorded.

The 3D particle coordinates are determined from this image sequence as follows: First, a 2D particle detection is performed for each image as described above. When the laser sheet is significantly thinner than the interparticle distance a particle is only visible in a single frame. Otherwise a particle appears in a short sequence of frames and reaches its maximum brightness when the particle is fully in the laser sheet which determines the depth position z of the particle.

Although this diagnostic setup is simple and allows to cover volumes of the order of several cm^3, the sequential image recording limits its use to static particle configurations. Nevertheless, scanning video microscopy has been successfully used to investigate crystalline states of 3D dust clouds [18–20]. An example of 3D structures measured with this scanning diagnostics is shown in Fig. 8.6.

A faster variant can be realized by moving the laser through the dust using rotating mirrors, see Fig. 10.5b. There only the laser is scanned through the dust cloud, the camera remains fixed. With the rotating mirrors the laser can be scanned at a quite high speeds allowing also to study dynamic processes. The scanning width through the dust cloud is limited by the depth of field[5] of the camera to about a few millimeters [28].

---

[5]The depth of field of a camera is the distance around the focal plane where the particles appear reasonably sharp. Large depths of field usually require small lens apertures.

10.5 Digital Holography

An interesting technique to encode 3D information in a single image is holography. One of the main advantages of digital holography is that this method does not require any lenses and is, in principle, not limited by the depth of field. It has been used in other research areas, see e.g. [29–31], and has been introduced to dusty plasmas by Kroll et al. [21, 32].

In this optical setup, see Fig. 10.6, a coherent planar light wave from a laser and a beam expander illuminates a dust particle (or an ensemble of particles). This is the so-called reference wave. A particle in the beam then scatters a (small) fraction of this light, the so-called object wave. Since both waves have a fixed phase relation they form an interference pattern that is recorded on the sensor of a camera without the use of any lenses. The typical interference pattern of a small dust particle is a Fresnel ring structure as indicated in Fig. 10.6. Fresnel rings consist of concentric dark and bright rings with varying radius. Figure 10.7a depicts a contrast-enhanced hologram of dust spheres with radius $a = 10 \pm 2\,\mu m$ inside an rf-discharge. The Fresnel rings of individual dust particles can be recognized with the help of the schematic drawing.

Now, from the holographic recording the three-dimensional particle positions have to be retrieved. This is done numerically based on Huygens principle: every point of the hologram is regarded as the origin of an elementary wave whose amplitude is defined by the intensity of the hologram at this point. The superposition of all these elementary waves forms the reconstructed object. The mathematical formulation of this principle is the Fresnel-Kirchhoff integral

$$U(x, y) = \int_{image} h(x', y') \frac{\exp(-2\pi i r/\lambda)}{r} dx' dy' \qquad (10.3)$$

with

$$r = \sqrt{(x' - x)^2 + (y' - y)^2 + d^2}. \qquad (10.4)$$

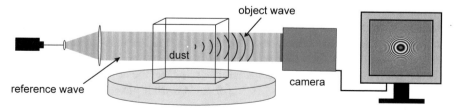

object wave

dust

reference wave

camera

Fig. 10.6 Schematic setup for digital in-line holography. A laser and a beam expander are used to illuminate the dust particles. The interference pattern of reference and object wave is recorded directly on the camera sensor. To avoid saturation of the sensor, a neutral density filter is mounted in front of the camera. From [27]

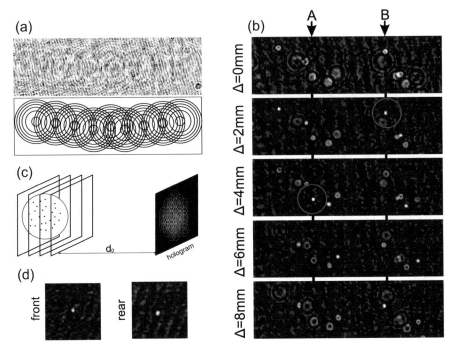

Fig. 10.7 Digital in-line holography of dusty plasmas: (**a**) Contrast enhanced hologram of dust particles in an rf-discharge (top). A sketch of the hologram is plotted below. (**b**) Reconstruction of the hologram in a distance $d_0 + \Delta$. Two particles (A, B) are labeled and their focus planes are marked with circles. (**c**) Illustration of the reconstruction geometry. The reconstruction planes shown in (**b**) are schematically indicated. The dust cloud is symbolized by the shaded sphere. (**d**) Focused particles with a distance of 40 mm between front and rear of a dust cloud to illustrate that digital in-line holography overcomes the usual problem of a limited depth of field. From [27]

Here, λ is the wavelength of the laser (wavelength of the reference wave) and $h(x', y')$ is the recorded hologram (with the image coordinates x', y'). The term $\exp(-2\pi i r/\lambda)/r$ represents the elementary spherical waves. Then, upon numerical solution of this integral [33], $U(x, y)$ is the reconstructed *complex* wave field where both the intensity and phase information are available [34].

The solution of Eq. (10.3) yields the complex wave field $U(x, y)$ at a certain distance d from the hologram plane. Thus, to reconstruct a 3D object, it is necessary to calculate $U(x, y)$ for successive values of d, each representing a reconstruction plane, see Fig. 10.7c. The distance between the reconstruction planes has to be chosen with respect to the desired spatial resolution and the object dimensions. For typical dusty plasma applications [21, 32] about 900–2000 planes with an interplane distance of 10 µm are required.

An example of reconstructed dust particle images is given in Fig. 10.7b. Here, the reconstruction planes at different distances $d_0 + \Delta$ are shown (here, $d_0 = 18$ cm corresponding to the approximate distance between the camera sensor and the dust cloud). The images show the reconstructed intensity $|U(x, y)|^2$. Two particles, A

and B are highlighted and marked by circles in their respective "focal" plane. Looking at the particles A and B in neighboring reconstruction planes, a clear change of particle shape and brightness is observed. The defocused particles appear ring-shaped and the reconstructed intensity decreases. Even though the original hologram in Fig. 10.7a has a quite low contrast, the signal to noise ratio of the reconstructed images is fully sufficient for a reliable particle detection.

However, the accuracy of determining the particle position along the optical axis d can be greatly enhanced by also considering the phase information, i.e. the imaginary part of $U(x, y)$. It is found [35], that the imaginary part of $U(x, y)$ vanishes at the "correct" depth d, where $|U(x, y)|^2$ has its maximum. Generally, the vanishing of the imaginary part of $U(x, y)$ can be detected with much higher accuracy than the maximum of the intensity $|U(x, y)|^2$.

Finally, Fig. 10.7d shows two focused particles at the rear and front of the dust cloud, 40 mm apart along the optical axis. This demonstrates the large depth of field of digital in-line holography. As a drawback, to achieve sufficiently strong interference patterns, relatively large (hollow glass) microspheres with diameters of the order of 20 μm are required.

An example of small reconstructed three-dimensional dust clusters is given in Fig. 10.8 where for comparison also top view images from a video microscope are shown. This example demonstrates the high accuracy in position determination using digital inline holography. Systems with larger number of particles (about 150) have been investigated, e.g. in [32].

10.6 Stereoscopy

The principal idea of stereoscopy is that a common observation volume is imaged by multiple cameras from different viewing angles. There are numerous applications of stereoscopy mimicking human vision, see e.g. [36]. For dusty plasmas, stereoscopic methods have been applied in different laboratories, see e.g. [15, 17, 22, 23, 37].

As an example, our stereoscopic camera setups in the laboratory and on parabolic flights [24, 38] are shown in Fig. 10.9. In both cases, four cameras image a common field of view. In our previous setups, only three cameras were used [10, 39, 40], but four cameras have been proven to provide by far superior data. The cameras are mounted at fixed positions with respect to each other. The cameras are synchronized to ensure that the frames of all cameras are recorded at the same instant.

To be able to retrieve the 3D particle positions from the camera images, in a first step the viewing geometry of the individual cameras has to be determined. Then, if the number of particles to be reconstructed is not too high, an algorithm with particle identification in the camera images and 3D reconstruction from identifying corresponding particles is used, the so-called multiset triangulation [23, 41, 42]. For denser dust clouds with higher particle number, the iterative shake-the-box-algorithm introduced for fluid systems by Schanz et al. [43] is employed. There, the 3D particle positions are determined by iterative matching of the observed camera images with artificial particle images based on the predicted 3D particle positions.

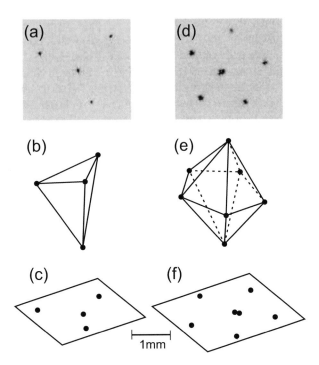

Fig. 10.8 Particle arrangement of small clusters ($N = 4$ and $N = 7$) measured simultaneously with digital holography and a video microscope (top view). (**a,d**) Inverted video microscope (top-view) images. (**b,e**) Average positions of the reconstructed clusters consisting of 4 (**b**) and 7 (**e**) particles. The marker size reflects the root mean squared displacement of particles from their equilibrium position. For comparison, the top-view projections are shown below (**c,f**). From [27]

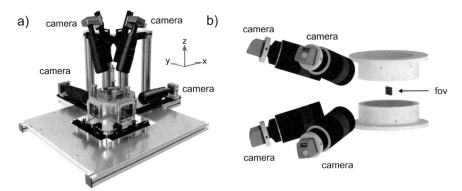

Fig. 10.9 (**a**) Stereoscopic imaging system used in the laboratory with four cameras: two side view cameras under 90° relative to each other and two top view cameras under a small relative angle. (**b**) Stereoscopic system used on parabolic flights with small relative angle. All four cameras observe the field of view (fov) through the same window of the discharge vessel (not shown). From [38]

The stereoscopic approach requires some technical effort in building a stable, reliable arrangement of multiple cameras. Also, sophisticated algorithmic and numerical methods are required to extract the three-dimensional particle positions. But stereoscopy allows to determine particle positions of several thousand particles in the field of view even at high particle density. The main limitation of stereoscopy is the depth of field of the cameras which usually is of the order of a few to some millimeters.

The individual technical steps to find the 3D particle positions will be discussed in the following.

10.6.1 Viewing Geometry

In a simple camera model [36] the viewing geometry of the camera can be described as follows. A point in the world coordinate system $\mathbf{M} = (X, Y, Z)^T$ is projected by the lens system onto a point $(x_p, y_p)^T$ on the camera image plane by

$$\begin{pmatrix} x'_p \\ y'_p \\ z'_p \end{pmatrix} = \mathsf{P} \cdot \begin{pmatrix} X \\ Y \\ Z \\ 1 \end{pmatrix} \quad \text{and} \quad \begin{pmatrix} x_p \\ y_p \end{pmatrix} = \begin{pmatrix} x'_p/z'_p \\ y'_p/z'_p \end{pmatrix}, \tag{10.5}$$

where P is the 3×4 projection matrix, see Fig. 10.10. This projection matrix

$$\mathsf{P} = \mathsf{K} \cdot [\mathsf{R}|\mathbf{t}] \tag{10.6}$$

contains the intrinsic 3×3 camera matrix K as well as the 3×3 rotation matrix R and the translation vector $\mathbf{t}$ that describe the camera orientation and position with

Fig. 10.10 Pin-hole camera model to illustrate the viewing properties

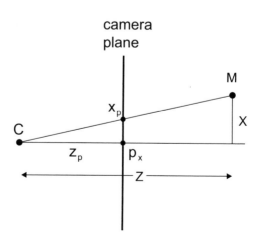

respect to the origin of the world coordinate system. The intrinsic camera matrix contains the camera-specific properties like focal length f and the principal point (image center) p and is then given by

$$K = \begin{pmatrix} f_x & 0 & p_x \\ 0 & f_y & p_y \\ 0 & 0 & 1 \end{pmatrix}. \tag{10.7}$$

From this it is determined how a point in the real world M is imaged onto the plane of the camera. In a stereoscopic camera system this projection matrix P has to be determined for each camera individually.

To determine the projection matrix for the cameras a calibration target with known properties is observed simultaneously with all cameras. Bouget [44] and Wengert et al. [45] have developed Matlab toolboxes that allow to reconstruct the projection matrices from the observation of a calibration target, see also [46]. For our dusty plasma experiments, a calibration target with a pattern of dots of 0.5 mm diameter on a 1 mm grid has been found to be useful (see Fig. 10.11). In addition, according to Wengert et al. [45], the target has two perpendicular bars in the center for unique identification of target orientation in the cameras.

This calibration target is moved and rotated in the field of view of all cameras and simultaneous images for the different target positions or orientations are captured. About 30 to 40 different target images per camera are sufficient to reliably determine the projection matrices P for each camera. For each image, the calibration dots are extracted using standard routines (as described above for particle identification), the central bars are identified from their ratio of major to minor axis as well as their absolute area [45]. The calibration dots are indexed by their position relative to the central bars. These coordinates are unique for every dot and are the same in all cameras and for all the different target orientations. This then allows to retrieve the line of sights of all cameras individually and the relative camera orientations, i.e. the camera projection matrices P [44, 46].

The calibration target with the perpendicular orientation bars together with the Matlab toolboxes [44–46] allow an automated analysis of the projection properties [47]. The camera calibration has to be performed each time the camera setup is changed (adjustment of focus, adjustment of mirrors etc.). In a parabolic flight campaign, the calibration is performed before and after each flight day. An accurate calibration will return more reliable 3D particle positions.

10.6.2   Multiset Triangulation

As mentioned above the multiset triangulation algorithm is an algorithm for retrieving 3D particle coordinates from multiple cameras. It is efficient if the particle number is not too high.

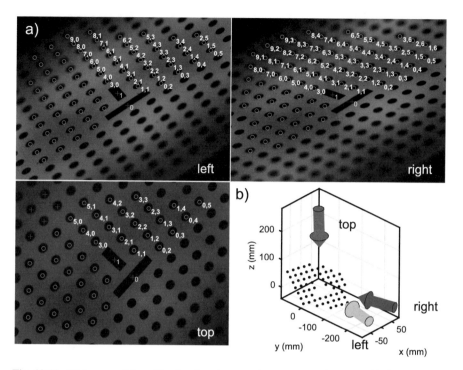

Fig. 10.11 (a) Images of the calibration target as seen in a stereoscopic setup with three cameras (one top view camera and the two side-view cameras). One sees the dot calibration pattern with the two bars in the center. For each view, processed data are overlaid, indicating the identified and indexed dots in the upper right quadrant. In the upper left quadrant only the identified dots without indexing are shown, indexing and identified dots have been omitted for the lower quadrants. (b) The lower right panel shows the camera positions and orientations reconstructed from the different views of the target. From [23]

There, the first step in data analysis is to identify the particles and their (2D) positions in the images of all cameras, see Sect. 10.2. Now, to find the 3D particle positions, it is necessary to identify corresponding particles in the different cameras ("stereo matching"), i.e. we have to determine which particle in camera C' is the same particle that we have observed in camera C. This is a difficult task since the particles all look the same, they are indistinguishable. In standard computer vision applications various image properties can be exploited to match corresponding items in the different images, such as color information, edges or patterns etc. All this does not work in our case of all particles represented by small patches of bright pixels against a darker background.

Epipolar Line Approach This problem can be solved when two (better three or even four or more) cameras are available. Then, corresponding particles can be identified from the corresponding epipolar lines [48]. The epipolar line l' is the line of sight of the image point $\mathbf{m}$ of camera C of a 3D point $\mathbf{M}$ as seen by camera C', see

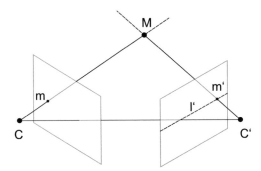

Fig. 10.12 Epipolar geometry for two cameras C and C′ facing the point **M**. The epipolar line l' is the line of sight from the image point **m** to the 3D point **M** of camera C as seen by camera C′

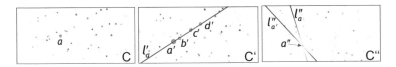

Fig. 10.13 Snapshot of a dust particle cloud from the three cameras of our stereoscopic setup together with epipolar lines to check for particle correspondences (the raw images are inverted: particles appear dark on a light background). See text for details. From [23]

Fig. 10.12. Here, **m** is the 2D position of a dust particle in the image of camera C and the point **M** corresponds to the real-world 3D coordinate of the particle, which is to be determined. Now, the corresponding projection **m′** of this point **M** in the image plane of the second camera C′ has to lie on the epipolar line l'.

This condition can be formulated as [36]

$$\tilde{\mathbf{m}}'^{T} \cdot \mathsf{F} \cdot \tilde{\mathbf{m}} = 0 \,, \tag{10.8}$$

where $\tilde{\mathbf{m}} = (m_x, m_y, 1)^{T}$ and F is the so called fundamental matrix

$$\mathsf{F} = \mathsf{K}'^{-T} [\mathbf{t}]_{\times} \mathsf{R} \mathsf{K}^{-1} \,, \tag{10.9}$$

which can be calculated from the camera and projection matrices of camera C and C′, see Eq. (10.6). Here, $[\mathbf{t}]_{\times}$ is the antisymmetric matrix such that $[\mathbf{t}]_{\times}\mathbf{x} = \mathbf{t} \times \mathbf{x}$.

Now, for a point **m** in camera C all possible candidates **m′** lying within a narrow stripe around the current epipolar line l' in C′ are taken as the *possible* corresponding particles to particle with image point **m**. Empirically, the stripe width around the epipolar line is chosen to be about 1 pixel in our case to account for inaccuracies in 2D particle position detection and calibration errors.

Figure 10.13 shows a snapshot of a dust particle cloud from three cameras of our stereoscopic setup. In the camera C a particle a is chosen and the corresponding epipolar line l'_a in the camera C' is calculated. In this case, there are 4 particles a'

to d' in this camera that have a small distance to the epipolar line and, thus, are possible corresponding particles to particle a. Now for the third camera C'', the epipolar line l_a'' of particle a from camera C in this camera C'' can be calculated as well as the epipolar lines of the particles a' to d' from C' in C''. Now, it is checked whether there is a particle in the camera C'' that is close to the epipolar line l_a'' of particle a and either of the epipolar lines of the possible candidates a' to d'. In this example, indeed a particle is found close to both the epipolar line l_a'' of particle a and the epipolar line $l_{a'}''$ of particle a'. Hence, here, a unique correspondence is found between particle a in camera C, a' in camera C' and a'' in camera C'' from a single snapshot. Having identified the particle correspondences the 3D position is determined from triangulation. An optimized triangulation procedure according to Hartley and Sturm [49] is usually used.

Multiset Information As seen above, we could identify a particle from the correspondence between the three cameras C, C' and C''. When now four (or more) cameras are available in the stereoscopic setup, there are multiple permutations of choosing C, C', C'' from the four (or more) available cameras. Hence, one now tries to find corresponding particles from all possible permutations of the cameras. When the found correspondence between the particles is the right one, one would expect to find real-world 3D positions of the particle that are very close in all the possible camera permutations.

Hence, a clustering algorithm is generally used to identify situations where all (or nearly all) camera permutations yield similar 3D particle positions. There, reconstructed positions from the different camera permutations that are clustered in a spatial region of a some microns in diameter are considered to be a single particle. When the number of reconstructed positions from the different permutations in a cluster exceeds a minimum number the particle is then taken as a real 3D particle. These particle position can be trusted since they are derived from a certain number of different camera permutations. Other particles that have been identified only in a smaller number of permutations might also be real particles, but are not further considered. In this way, also images where particles might overlap and/or a particle might be occluded by another in one of the cameras still leads to identification of reliable particles.

Trajectory Linking The tracking algorithm in 3D is essentially identical to the 2D situation described above: based on a Kalman filter approach the 3D location of a particle in the next frame is predicted (and corrected) from the trajectory in the past and present frame.

10.6.3 Shake-the-Box-Algorithm

When a large number of particles is seen in the field of view and their 3D positions are to be retrieved the multiset triangulation becomes inefficient as the computing time would be too exhaustive and many erroneous detections would impact the analysis quality.

Here, iterative reconstruction algorithms, such as the shake-the-box-algorithm [43], have the capability to evaluate the trajectories of a large number of particles. This algorithm has been developed by Schanz et al. [43] to follow tracer particles in fluid systems, but has been successfully implemented for dusty plasmas [24, 46].

The general idea is that the predicted particle position is iteratively matched to the observed particle position directly in the camera images. The matching criterion is based on a comparison of the observed camera images and calculated artificial particle images. The following procedure illustrates the main steps.

First of all, the cameras have to be calibrated, i.e. the projection matrices P of the cameras have to be determined as described above (Sect. 10.6.1). The initial trajectories in the first few frames to start the shake-the box iteration are obtained using, e.g. the above mentioned multiset triangulation (Sect. 10.6.2).

The shake-the-box algorithm then starts with the particle positions X_p as predicted by the Kalman filter from the trajectories in the first frames. These predicted positions X_p are used to generate an artificial projected particle image for each camera. A particle in the computed image is modeled, e.g., by a two-dimensional Gaussian intensity profile in a 10×10 pixels window. The width of the Gaussian typically is a few pixels and must be adapted to the approximate particle size in each recorded image. Next, the particle position X_p is artificially moved by an amount Δ along the 3D axes ("shaking") as sketched by Fig. 10.14a. For each shaking step an artificial image subset is calculated where the 2D particle image of the particle at the 3D position $X_p + \Delta$ is determined in all cameras using the projection matrices P, see Fig. 10.14b.

Then, the difference between the artificial computed image and the measured image is computed for all cameras as $\sum_k I_{m,k} - I_{c,k}$, where $I_{m,k}$ is the pixel intensity of pixel k of the measured image and $I_{c,k}$ that of the computed image. The assumed particle position with the lowest residue will be used as the actual position in the next shaking step in this iterative process. It is reasonable to start with two coarse

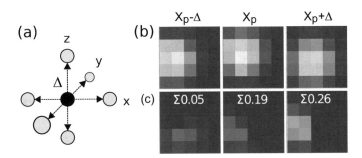

Fig. 10.14 (**a**) "Shaking" the particle means to shift the position by a given step size Δ along the x-, y- and z-axis. (**b**) Example showing calculated particle images at the current position X_p and shifted along x to $X_p \pm \Delta$. (**c**) Residue between calculated and measured image where the value of the residue is indicated. In this step, $X_p - \Delta$ would be the position for the next iteration. From [24]

shaking iterations (with $\Delta \approx 15\,\mu m$) followed by eight to ten fine shaking iterations ($\Delta \approx 7\,\mu m$). The particle position has converged when it results in the smallest residue between calculated and measured image in all cameras. This position is then added to the trajectory. The trajectory is discontinued when either the intensity of a particle in the image falls below a given threshold in one of the cameras (since, e.g. there is no particle at the projected location in the recorded image) or the iterative shaking process has not converged after a maximum number of allowed iterations (typically 8 to 10).

After the shaking iterations, the predicted positions have been either confirmed and optimized or declined. For a second cycle, all confirmed particles are projected to build an artificial image which is then subtracted from the measured image. This difference image now contains mainly particles that have newly entered the field of view or particles whose trajectory has been discontinued prematurely. It is then processed again by multiset triangulation algorithms to find the positions $\mathbf{X}_p$ of the remaining visible particles. These particles then undergo a usual shaking optimization cycle. Most, but not all, particle positions can be successfully reconstructed using the shake-the-box algorithm.

The stereoscopic setup with 4 cameras indicated in Fig. 10.9b has been used on parabolic flights. The cameras had a common field-of-view of about $16 \times 13\,mm^2$ in the central plane of the discharge. A laser is expanded into a light sheet of 2 mm width which defines the depth of the three-dimensional volume for particle tracking. Snap shots of the dust cloud as viewed by two of the four cameras are shown in Fig. 10.15a.

From the video sequences of the 4 cameras the trajectories within the dust cloud have been obtained using the shake-the-box-algorithm [24]. In this example, Fig. 10.15b, up to 4500 particles have been reconstructed in each frame. A total number of 21,000 trajectories have been obtained. One can easily identify the

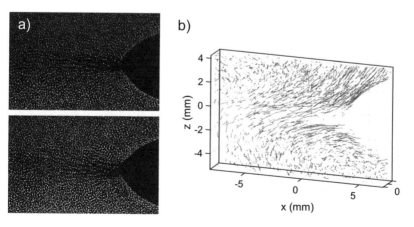

Fig. 10.15 (a) Snap shot of the dust cloud viewed by camera 1 and camera 2 of the four-camera setup shown in Fig. 10.9b. (b) Three-dimensional view of the particle trajectories in the field of view. For clarity only 10% of the reconstructed trajectories are shown. After [24]

motion of the particles along the void boundary and the lanes where the particles perform mainly an outward motion. In the upper and lower parts of the cloud one can see the a vortex-like motion. This vortex-like motion is driven by gradients of the ion streaming motion [50].

10.7   Imaging Diagnostics for Nanoparticles

Plasmas with nanoparticles can be formed in discharges with reactive gases, as discussed in Sect. 11.1. Such plasmas produce volume-filling clouds with a high density of dust particles in the nanometer size range. This might lead to interesting changes of the plasma properties, e.g. by electron depletion (see Sect. 2.6 and also [51]). For a good description of the particle influence on the plasma a reliable, spatially and temporally resolved, dust size and dust density diagnostic is required.

10.7.1   Dust Size Diagnostics

For an imaging dust size diagnostics the light scattering properties of (nano-) particles are exploited. Light scattering has generally been discussed above in Sect. 10.1. For nanoparticles, one makes use of the fact that the light is scattered differently by a particle when the incident light is polarized in the scattering plane ($I_{\parallel}$) compared to light polarized perpendicular to the plane ($I_{\perp}$), see e.g. [1]. Hence, the ratio of $I_{\parallel}/I_{\perp}$, or alternatively, the scattering parameter

$$q = \frac{I_{\parallel} - I_{\perp}}{I_{\parallel} + I_{\perp}} , \qquad (10.10)$$

can serve as a measure of the particle size at a fixed observation angle, see Fig. 10.16.

Here, we like to point out that, at optical wavelengths, the polarization degree of light scattered by the particles can be used as a size diagnostic for particles in a typical size range between 50 to 200 nm (depending on the index of refraction of the dust material). For smaller particles, the scattering intensity is too low, for larger particles the strong oscillations of the scattering intensity with varying dust size leads to non-unique relations between polarization degree and dust size, see Fig. 10.16. Also, it should be noted that the entire discussion of this section relies on single light-scattering events of individual particles. Multiple scattering is not taken into account, here, see also [52].

The behavior of the scattering parameter q can be used to setup an imaging size diagnostic where the dust size can be measured spatially and temporally resolved by measuring both polarization states $I_{\parallel}$ and $I_{\perp}$ simultaneously, see Fig. 10.17a and [53]. For that purpose the incident light is linearly polarized at $45°$ with respect to the scattering plane, so that the incident light has the same intensity in both

Fig. 10.16 Ratio of the scattering intensity of a dust particle with parallel and perpendicular polarization of the incident light, $I_{\parallel}/I_{\perp}$ under an angle of 90°. The inset shows the polarization parameter $q = (I_{\parallel} - I_{\perp})/(I_{\parallel} + I_{\perp})$. This shows that the polarization degree can be used for a certain size range. Other parameters: light wavelength 532 nm, index of refraction 1.68+0.03i

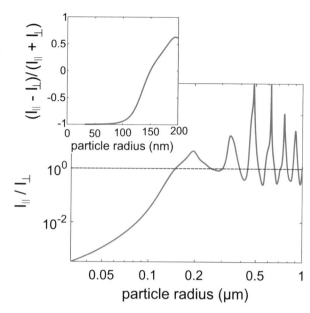

polarization states. This light illuminates a slice of a dust cloud. The scattered light is then recorded under a fixed angle of 90° in this case. Two CCDs are mounted behind a polarizing beam splitter, so that both cameras observe the same region of interest, one camera in the parallel-polarized light and the other in the perpendicular-polarized light. Then, for each pixel (or group of pixels) the scattering parameter q is derived from the light intensity measured by the cameras. However, unwanted contributions of unpolarized light, e.g. from multiple scattering, plasma glow etc., measured by the CCDs may interfere with the size measurement.

An example of such a measurement is shown in Fig. 10.17b where the spatially resolved q-parameter (and the corresponding dust size) is shown for different instants during a growth cycle in an acetylene plasma. First, the dust cloud consists of small particles. Also, the central void is clearly seen. Then, a quite homogeneous particle growth throughout the cloud is found. At later stages, "growth fronts" are found to propagate through the dust cloud. Also, it is seen that the larger particles sediment to the lower parts of the cloud. This continues until the particles are so large that they are expelled from the plasma volume by the action of various forces. Then, another growth cycle can start. This example demonstrates the capability of this imaging approach [51].

Alternatively, especially for particle sizes in the Mie regime, one can exploit the angular dependence of the scattering intensity, see also Sect. 10.1 or [54, 55]. There, we have seen that for particles with a size of the order of the laser wavelength the scattering intensity has a distinct angular distribution that depends on the dust size (and the index of refraction). This angular scattering behavior can then be used as a size diagnostics. In corresponding experiments [55], particles have been introduced

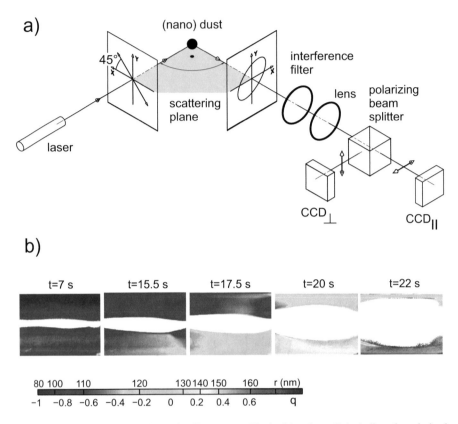

Fig. 10.17 (a) Setup of imaging Mie ellipsometry. The incident laser light is linearly polarized with an angle of 45° with respect to the plane of incidence. The scattered light is observed under a scattering angle of 90°, split into its parallel and perpendicular part by a polarizing beam splitter and detected by two CCD cameras, $CCD_{\parallel}$ and $CCD_{\perp}$. (b) Spatiotemporal evolution of the q-parameter during one growth cycle. The color corresponds to the radius of the growing particles in the illuminated cross section of the dust cloud. The dust cloud fills almost the entire space between the electrodes, the central void is clearly seen. Sedimentation and several "growth fronts" are observed during the growth of the particles. After [53], figure by courtesy of Greiner

into the discharge and the camera has been rotated around a slice of a dust cloud illuminated by a laser, see Fig. 10.18a. The angle-dependent scattering intensity of the different parts of the dust cloud has been measured spatially resolved. In Fig. 10.18b,c the angular-resolved scattering intensity is shown for a group of pixels representing a small fraction of the dust cloud. For comparison, the Mie scattering signal of a particle of 1.5509 μm radius is shown along with the measurement. Due to the strong angle and size-dependent oscillations of the scattering intensity, the comparison allows a very accurate size determination, the uncertainty of the measured dust size is of the order of only 1%.

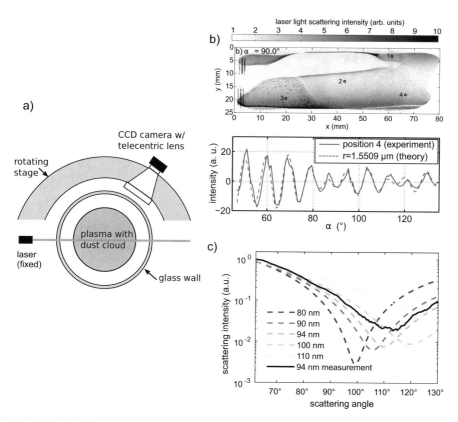

Fig. 10.18 (a) Sketch of the diagnostic setup for angle-resolved Mie scattering (top view). (b) Image of a typical dust cloud of plastic particles of about 1.5 μm radius under a scattering angle of 90.0° and angular resolved scattering intensity at point 4. The measured scattering intensity is shown together with a respective fit from Mie scattering theory. (c) Same as (b) for a dust cloud consisting of submicron Al₂O₃ particles. The solid line corresponds to the measurement, the dashed lines denote the calculated Mie scattering for different particle sizes. From [55, 56]

However, for particles in the submicron range that are smaller than the laser wavelength the angular scattering intensity distribution is much smoother and nearly featureless. The size can still be measured from the angular distribution, but the uncertainty is more of the order of 10%, see Fig. 10.18c.

10.7.2 Dust Density Diagnostics

The dust density can be measured by an absorption measurement exploiting Beer-Lambert's law

$$I = I_0 \exp\left(-\langle n_{\mathrm{d}}\sigma\rangle L\right) . \tag{10.11}$$

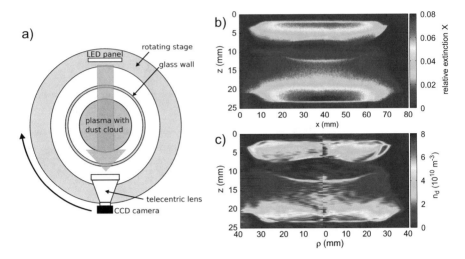

Fig. 10.19 (**a**) Scheme of the experimental setup for the density measurements. (**b**) Measured absorption I/I_0 yielding the line-integrated absorption for micrometer dust. (**c**) Radial density distribution reconstructed from the Abel inversion of the absorption. From [57]

Here, one measures the intensity I of a light source transilluminating the dust cloud in comparison to the light intensity I_0 without dust. Then, the dust density n_d can be derived from the absorption cross section σ and the length of the light path through the cloud L. The cross section σ can be determined from the Mie scattering properties of the particles.

In the experimental realization [57], a planar light source, such as an LED panel, is used to transilluminate the dust cloud. The transmitted light is recorded via a camera equipped with a telecentric lens, see Fig. 10.19a. The telecentric lens ensures that parallel light rays are imaged onto the camera chip. Then by comparison with a plasma without dust cloud the light intensities I and I_0 are evaluated for each pixel on the camera chip (or for regions consisting of a few pixels) leading to an absorption $\langle n_d\sigma\rangle$ that is spatially resolved in the image plane. However, with this technique, the measured quantity $\langle n_d\sigma\rangle$ is determined averaged along the line-of-sight, i.e.,

$$\langle n_d\sigma\rangle(x, z) = \frac{1}{L} \int_{-L/2}^{L/2} n_d(\varrho, z)\sigma \, dy \,. \tag{10.12}$$

Here, x, z denote the coordinates in the image plane, and y denotes the direction of the light path and ϱ is the cylindrical radius $\varrho = (x^2 + y^2)^{1/2}$.

Under the assumption that the dust cloud has an axial symmetry the absorption in the image plane can be transformed into the radially and vertically resolved absorption by Abel inversion

$$n_d(\varrho, z) = -\frac{1}{\pi} \int_\varrho^\infty \frac{d}{dx} n_d(x, z) \frac{dx}{\sqrt{x^2 - \varrho^2}}, \qquad (10.13)$$

which can be solved by numerical techniques [57–60].

As an example, we show the spatially resolved intensity ratio $I/I_0(x, z)$ as measured from two camera images with and without dust cloud, see Fig. 10.19b. This intensity ratio reflects the line-integrated absorption $n_d(x, z)$. We also show the Abel-inverted intensity density $n_d(\varrho, z)$. In this case, a cloud of monodisperse particles of 3.55 μm diameter has been used. Hence, the scattering cross section is constant (here, $\sigma = 2.2 \times 10^{-11}$ m^2 from Mie scattering [1]).

The line-integrated absorption and its Abel inversion, Fig. 10.19b,c are similar, but it is seen that the line-integrated absorption reaches its maximum near the central axis (due to the longer absorption pathways, there). However, the Abel inversion demonstrates that the highest density is found in the outer parts of the dust cloud [57, 61].

When an axial symmetry of the dust cloud cannot be guaranteed the dust density can be retrieved from computed tomography techniques, see [57] which, however, requires to measure the light absorption along various directions.

References

1. H. van de Hulst, *Light Scattering by Small Particles* (Dover, New York, 1981)
2. H. Nobach, M. Honkanen, Exp. Fluids **38**, 511 (2005)
3. V. Nosenko, J. Goree, A. Piel, Phys. Plasmas **13**, 032106 (2006)
4. Y. Ivanov, A. Melzer, Rev. Sci. Instrum. **78**, 033506 (2007)
5. Y. Feng, J. Goree, B. Liu, Rev. Sci. Instrum. **78**, 053704 (2007)
6. J.C. Crocker, D.G. Grier, J. Colloid Interface Sci. **179**, 298 (1996)
7. D. Blair, E. Dufresne, *The Matlab Particle Tracking Code Repository.* http://site.physics.georgetown.edu/matlab/
8. R.S. Bucy, B.G. Williams, *Lectures on Discrete Time Filtering. Signal Processing and Digital Filtering* (Springer, New York, 1994)
9. A. Savitzky, M.J.E. Golay, Anal. Chem. **36**(8), 1627 (1964). http://dx.doi.org/10.1021/ac60214a047
10. B. Buttenschön, M. Himpel, A. Melzer, New J. Phys. **13**, 023042 (2011)
11. M. Raffel, C. Willert, J. Kompenhans, *Particle Image Velocimetry: A Practical Guide* (Springer Verlag, Berlin, 1998)
12. T. Miksch, A. Melzer, Phys. Rev. E **75**, 016404 (2007)
13. E. Thomas, Phys. Plasmas **8**, 329 (2001). http://dx.doi.org/10.1063/1.1328355
14. E. Thomas, Phys. Plasmas **9**, 17 (2002)
15. E. Thomas, J.D. Williams, J. Silver, Phys. Plasmas **11**, L37 (2004)
16. E. Thomas, J. Williams, Phys. Plasmas **13**(5), 055702 (2006). http://dx.doi.org/10.1063/1.2174831
17. J.D. Williams, Phys. Plasmas **18**, 050702 (2011)

18. J. Pieper, J. Goree, R. Quinn, J. Vac. Sci. Technol. A **14**, 519 (1996)
19. M. Zuzic, A.V. Ivlev, J. Goree, G.E. Morfill, H.M. Thomas, H. Rothermel, U. Konopka, R. Sütterlin, D.D. Goldbeck, Phys. Rev. Lett. **85**, 4064 (2000)
20. O. Arp, D. Block, A. Piel, A. Melzer, Phys. Rev. Lett. **93**, 165004 (2004)
21. M. Kroll, D. Block, A. Piel, Phys. Plasmas **15**, 063703 (2008)
22. S. Käding, A. Melzer, Phys. Plasmas **13**, 090701 (2006)
23. A. Melzer, M. Himpel, C. Killer, M. Mulsow, J. Plasma Phys. **82**, 615820102 (2016). http://dx.doi.org/10.1017/S002237781600009X, http://journals.cambridge.org/article_S002237781600009X
24. M. Himpel, S. Schütt, W.J. Miloch, A. Melzer, Phys. Plasmas **25**(8), 083707 (2018). http://dx.doi.org/10.1063/1.5046049
25. B.M. Annaratone, T. Antonova, D.D. Goldbeck, H.M. Thomas, G.E. Morfill, Plasma Phys. Control. Fusion **46**, B495 (2004)
26. P. Hartmann, I. Donko, Z. Donko, Rev. Sci. Instrum. **84**, 023501 (2013). http://dx.doi.org/10.1063/1.4789770, http://scitation.aip.org/content/aip/journal/rsi/84/2/10.1063/1.4789770
27. D. Block, A. Melzer, in *Introduction to Complex Plasmas*, ed. by M. Bonitz, N. Horing, P. Ludwig, Springer Series on Atomic, Optical, and Plasma Physics (Springer, New York, 2010), pp. 135–154
28. D. Samsonov, A. Elsaesser, A. Edwards, H.M. Thomas, G.E. Morfill, Rev. Sci. Instrum. **79**, 035102 (2008)
29. K.D. Hinsch, Meas. Sci. Technol. **13**(7), R61 (2002). http://stacks.iop.org/0957-0233/13/i=7/a=201
30. F. Dubois, L. Joannes, J.C. Legros, Appl. Opt. **38**(34), 7085 (1999). http://dx.doi.org/10.1364/AO.38.007085. http://ao.osa.org/abstract.cfm?URI=ao-38-34-7085
31. G. Indebetouw, W. Zhong, J. Opt. Soc. Am. A **23**, 1699 (2006). http://dx.doi.org/10.1364/JOSAA.23.001699. http://josaa.osa.org/abstract.cfm?URI=josaa-23-7-1699
32. M. Kroll, J. Schablinski, D. Block, A. Piel, Phys. Plasmas **17**, 013702 (2010)
33. T.H. Demetrakopoulos, R. Mittra, Appl. Opt. **13**(3), 665 (1974). http://dx.doi.org/10.1364/AO.13.000665. http://ao.osa.org/abstract.cfm?URI=ao-13-3-665
34. J. Garcia-Sucerquia, W. Xu, S.K. Jericho, P. Klages, M.H. Jericho, H.J. Kreuzer, Appl. Opt. **45**, 836 (2006). http://dx.doi.org/10.1364/AO.45.000836. http://ao.osa.org/abstract.cfm?URI=ao-45-5-836
35. G. Pan, H. Meng, Appl. Opt. **42**, 827 (2003). http://dx.doi.org/10.1364/AO.42.000827. http://ao.osa.org/abstract.cfm?URI=ao-42-5-827
36. R.I. Hartley, A. Zisserman, *Multiple View Geometry in Computer Vision*, 2nd edn. (Cambridge University Press, Cambridge, 2004)
37. T. Antonova, B.M. Annaratone, D.D. Goldbeck, V. Yaroshenko, H.M. Thomas, G.E. Morfill, Phys. Rev. Lett. **96**, 115001 (2006)
38. M. Mulsow, M. Himpel, A. Melzer, Phys. Plasmas **24**(12), 123704 (2017). http://dx.doi.org/10.1063/1.5006841
39. M. Himpel, C. Killer, B. Buttenschön, A. Melzer, Phys. Plasmas **19**, 123704 (2012)
40. M. Himpel, C. Killer, A. Melzer, T. Bockwoldt, K.O. Menzel, A. Piel, Phys. Plasmas **21**, 033703 (2014)
41. Y. Akhmetbekov, V. Lozhkin, D. Markovich, M. Tokarev, in *9th International Symposium on Particle Image Velocimetry-PIV*, vol. 11 (2011), pp. 21–23
42. B. Wieneke, Exp. Fluids **45**, 549 (2008). http://dx.doi.org/10.1007/s00348-008-0521-5
43. D. Schanz, S. Gesemann, A. Schröder, Exp. Fluids **57**, 70 (2016). http://dx.doi.org/10.1007/s00348-016-2157-1.
44. J.Y. Bouguet, Camera calibration toolbox for matlab (2008). http://www.vision.caltech.edu/bouguetj/calib_doc/index.html
45. C. Wengert, M. Reeff, P.C. Cattin, G. Székely, in *Bildverarbeitung für die Medizin* (Springer-Verlag, Berlin, 2006), pp. 419–23. ftp://ftp.vision.ee.ethz.ch/publications/proceedings/eth_biwi_00381.pdf

46. M. Himpel. Camera calibration and 3d particle tracking toolbox for matlab (2016). https://physik.uni-greifswald.de/ag-melzer/forschung/stereo-toolbox
47. M. Himpel, B. Buttenschön, A. Melzer, Rev. Sci. Instrum. **82**, 053706 (2011)
48. Z. Zhang, Int. J. Comput. Vis. **27**, 161 (1998). http://dx.doi.org/10.1023/A%3A1007941100561
49. R.I. Hartley, P. Sturm, Comput. Vis. Image Underst. **68**, 146 (1997). http://dx.doi.org/10.1006/cviu.1997.0547. http://www.sciencedirect.com/science/article/pii/S1077314297905476
50. T. Bockwoldt, O. Arp, K.O. Menzel, A. Piel, Phys. Plasmas (1994–Present) **21**, 103703 (2014)
51. F. Greiner, A. Melzer, B. Tadsen, S. Groth, C. Killer, F. Kirchschlager, F. Wieben, I. Pilch, H. Krüger, D. Block, A. Piel, S. Wolf, Eur. Phys. J. D **72**, 81 (2018). https://doi.org/10.1140/epjd/e2017-80400-7
52. F. Kirchschlager, S. Wolf, F. Greiner, S. Groth, A. Labdon, Appl. Phys. Lett. **110**(17), 173106 (2017). http://dx.doi.org/10.1063/1.4982645
53. F. Greiner, J. Carstensen, N. Köhler, I. Pilch, H. Ketelsen, S. Knist, A. Piel, Plasma Sources Sci. Technol. **21**(6), 065005 (2012). http://stacks.iop.org/0963-0252/21/i=6/a=065005
54. W.W. Stoffels, E. Stoffels, G. Swinkels, M. Boufnichel, G. Kroesen, Phys. Rev. E **59**, 2302 (1999)
55. C. Killer, M. Mulsow, A. Melzer, Plasma Sources Sci. Technol. **24**, 025029 (2015). http://stacks.iop.org/0963-0252/24/i=2/a=025029
56. H. Krüger, C. Killer, S. Schütt, A. Melzer, Plasma Sources Sci. Technol. **27**, 025004 (2018). http://iopscience.iop.org/article/10.1088/1361-6595/aaa7d5
57. C. Killer, M. Himpel, A. Melzer, Rev. Sci. Instrum. **85**, 103711 (2014)
58. G. Pretzler, Z. Naturforsch. **46a**, 639 (1991)
59. C. Killer, F. Greiner, S. Groth, B. Tadsen, A. Melzer, Plasma Sources Sci. Technol. **25**, 055004 (2016). http://dx.doi.org/10.1088/0963-0252/25/5/055004. http://stacks.iop.org/0963-0252/25/i=5/a=055004
60. C. Killer. Abel inversion algorithm, matlab central file exchange (2013). https://www.mathworks.com/matlabcentral/fileexchange/43639-abel-inversion-algorithm
61. A. Melzer, M. Himpel, H. Krüger, S. Schütt, Plasma Phys. Controlled Fusion **61**, 014029 (2018). https://doi.org/10.1088/1361-6587/aad652

Chapter 11
Particle Growth in Dusty Plasmas and Applications

After discussing various fundamental aspects of dusty plasmas this chapter will deal with more "applied" questions. Here, particle growth mechanisms in low-temperature plasmas are presented and the role of grown dust in processing plasmas is briefly discussed in view of the formation of "new" materials, dust removal and dust in fusion devices. For a more detailed overview the reader is referred to, e.g. [1–5]. Here, only a gross exemplary overview is intended.

11.1 Particle Growth Mechanisms

Many technological plasmas are operated in reactive gases. Typical examples are silicon-containing gases like silane (SiH_4), or organo-silicons like hexamethyldis-iloxane (HMDSO, $O[Si(CH_3)_3]_2$). Also, hydrocarbons (such as methane, CH_4, and acetylene, C_2H_2), and fluorocarbons (e.g. carbon tetrafluoride CF_4) are frequently used. These gases are required to deposit thin films on substrates or to etch into silicon layers, photoresists or protective films. For example, silane is used to deposit thin (amorphous) silicon films. This is needed in computer chip manufacturing or in solar cell production making use of the properties of silicon as a semiconductor. Hydrocarbons are used, e.g., to form diamond-like carbon films. There it is intended to grow carbon films that have the properties of diamond, especially in view of hardness. These films are used as covering layers to resist against wear. In addition, HMDSO is used as barrier layers in PET bottles or Tetra packs. HMDSO provides SiO_2-layers that prevent the carbon-dioxide in beer, lemonade etc. to diffuse through the walls of the container.

All these reactive gases have the tendency to polymerize under plasma conditions. The particle growth mechanisms will be discussed in some more detail in the following. As a model system, we will study the growth mechanism in silane, since it is one of the most investigated systems and is of technological relevance.

© Springer Nature Switzerland AG 2019
A. Melzer, *Physics of Dusty Plasmas*, Lecture Notes in Physics 962,
https://doi.org/10.1007/978-3-030-20260-6_11

Fig. 11.1 Schematic
representation of the
evolution of particle size and
particle density in a silane
discharge as a function of
time after discharge ignition.
After [6, 7]. The growth steps
are also schematically
indicated, after [4]

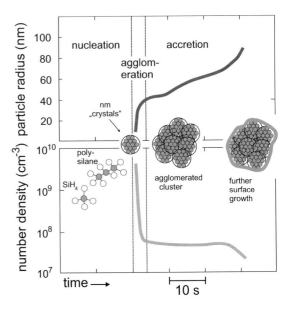

Silane, SiH_4, has the same molecular structure as methane (CH_4), where the 4 hydrogen atoms are on the edges of a tetrahedron and the silicon atom is sitting in the center. The case of silane can provide a general insight into the mechanisms of particle growth. Other reactive gases certainly behave differently due to the different chemical properties and reactions involved, but silane shall serve as a paradigm here.

In a plasma discharge (typically rf discharge), the particles grow from molecules (of Å size) to a few hundred nanometer in diameter on the time scale of seconds, see Fig. 11.1. The formation of dust particles generally works best for high discharge powers and higher gas pressure. The growth is not homogeneous, but one can identify different phases of particle growth, see [1, 5, 8]:

1. **Cluster formation and nucleation:** The particles grow from molecules to molecular clusters of a few nanometer in size.[1] This phase is dominated by plasma chemistry.
2. **Agglomeration or coagulation phase:** The particles grow rapidly from a few nanometer to 50 nanometer, say. Correspondingly, the density of particles drops dramatically.
3. **Accretion phase:** One finds a slow increase in particle size and roughly constant particle density. This phase is dominated by powder dynamics.

The coagulation and accretion phase are illustrated in Fig. 11.1 where the temporal evolution of dust size and dust density are shown [6]. In the coagulation

---

[1]In this chapter, clusters are seen as molecular clusters consisting of a certain number of atoms and molecules in contrast to the Yukawa clusters in the previous chapters that denoted an ensemble of dust particles.

phase, around some ten seconds after discharge ignition, one sees a rapid growth of the dust size accompanied by a dramatic decrease in density. In the accretion phase a continuous growth at nearly constant density is found. These measurements have been obtained using a sophisticated laser light scattering technique which is sensitive to particle sizes of down to about ten nanometer. The initial growth phase with molecular cluster formation, is not resolved, here, and has to be addressed by different methods, e.g. mass spectrometry. In the following, we will describe the different phases in some more detail.

11.1.1 Cluster Formation

In the cluster formation phase, the dust particles grow from molecules to molecular clusters of a few nanometer in size. Nanometer sized particles contain of the order of 100 atoms or molecules. For such small particles (molecules), chemical processes dominate the growth. Possible chemical reactions in the case of silane can be summarized in the form

$$Si_n H_m + SiH_4 \longrightarrow Si_{n+1} H_{m'} + (H, H_2) \text{ products} \tag{11.1}$$

$$Si_n H_m^+ + SiH_4 \longrightarrow Si_{n+1} H_{m'}^+ + (H, H_2) \text{ products} \tag{11.2}$$

$$Si_n H_m^- + SiH_4 \longrightarrow Si_{n+1} H_{m'}^- + (H, H_2) \text{ products} \tag{11.3}$$

where reaction pathways are described for neutral, positively charged and negatively charged molecules, respectively.

In this phase, the particles can be measured by mass spectrometry. With modern mass spectrometers, particle masses up to 2000 amu can be easily measured with high temporal resolution [9–11]. That means, that clusters with about 60 Si-atoms can be identified with such a device (Si has a mass of 28 u). Additionally, mass spectra of positively charged particles (ions), negative ions or neutrals can be discriminated by applying opposite voltages at the entrance orifice of the mass spectrometer to repel the unwanted species. The result of such a measurement for molecular clusters up to 20 silicon atoms is shown in Fig. 11.2.

One can easily see that negatively charged molecular clusters are easily found with up to 20 silicon atoms (and also beyond), whereas positive clusters can be seen only up to 13 or 14 silicon atoms. Neutral clusters extend only to molecules with about 4 silicon atoms. This finding clearly suggests that the reaction pathway is dominated by negative ions. This can be understood from the fact that the plasma potential is typically positive with respect to the walls. Thus, negative particles, like electrons and the negative silane clusters are trapped and confined to the plasma. Positive clusters are readily driven to the walls or electrodes. Thus negative molecules have longer residence times in the plasma and can thus dominate the

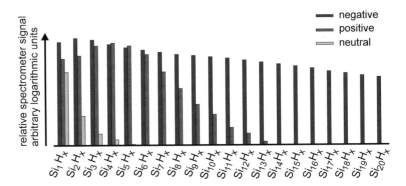

Fig. 11.2 Relative abundance of negative, positive ions and neutral molecules from mass spectrometry in a silane discharge up to clusters containing 20 silicon atoms (with an atomic mass of about 600). Schematic representation after [11]

cluster phase. A further and important reason seems to be the fact that the reaction chain with negative molecules is also favored from chemistry.

In view of the negative clusters the favored so-called sylil reaction pathway is

$$\text{SiH}_3^- + \text{SiH}_4 \longrightarrow \text{Si}_2\text{H}_5^- + \text{H}_2 \tag{11.4}$$

$$\text{Si}_2\text{H}_5^- + \text{SiH}_4 \longrightarrow \text{Si}_3\text{H}_7^- + \text{H}_2 \tag{11.5}$$

$$\cdots$$

$$\text{Si}_n\text{H}_{2n+1}^- + \text{SiH}_4 \longrightarrow \text{Si}_{n+1}\text{H}_{2n+3}^- + \text{H}_2 \tag{11.6}$$

where SiH_3^- is the precursor of this reaction chain. This ion is formed by dissociative attachment, i.e. an electron attaches to the SiH_4 molecule and a hydrogen ion is removed from the molecule, namely

$$\text{SiH}_4 + \text{e}^- \longrightarrow \text{SiH}_3^- + \text{H}.$$

Following this reaction chain, the ratio of hydrogen to silicon in a cluster then should be $[\text{H}]:[\text{Si}] = (2n + 1)/n$ which is indeed found from the mass spectra for small Si-clusters with up to $n = 5$ or 6 silicon atoms. For large silicon clusters with $n > 10$ the mass spectra show a concentration of hydrogen to silicon that is very close to $[\text{H}]:[\text{Si}] = 4:3$. This means that there are equal number of Si–Si and Si–H bonds. Thus, at free bonds randomly hydrogen or silicon is attached. From infrared absorption spectroscopy one knows that there are no double bonds in Si clusters.

Another possible pathway contains even numbers of hydrogen atoms in the cluster according to the reaction (so-called silylene pathway) [5]

$$\text{Si}_n\text{H}_{2n}^- + \text{SiH}_4 \longrightarrow \text{Si}_{n+1}\text{H}_{2n+2}^- + \text{H}_2.$$

Fig. 11.3 Examples of
structures of small
hydrogenated Si$_n$ clusters for
$n = 1$ to 13. Note the
formation of rings for $n \geq 3$
and "cages" for $n > 10$. After
[12, 13]

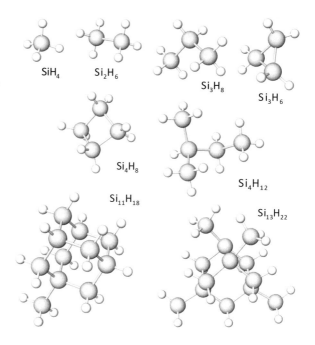

The structures of various hydrogenated silicon clusters Si$_n$H$_m$ are shown in
Fig. 11.3 as found from [12, 13]. The structure of small clusters ($n < 4$) form silyls
or silylenes. Larger clusters $n > 4$ can form rings or more complicated structures.
The hydrogen-to-silicon ratio drops accordingly from ≈ 2 to smaller values. Since
the Si–Si bond length is about 2.5 Å, about $n = 60$ silicon atoms would be needed
to form a particle of 1 nm diameter, since $\sqrt[3]{60} \times 2.5$ Å ≈ 1 nm.

11.1.2 Agglomeration/Coagulation Phase

In the agglomeration phase, the particles rapidly grow from a few nanometer to
several ten nanometers. At the same time, the density of clusters dramatically
drops by several orders of magnitude. However, it can be estimated from the
measurements (see Fig. 11.1) that the overall *volume* of the clusters does not change
very drastically. From that, it is induced that the large number of small particles
agglomerate (stick together) to form a smaller number of larger molecular clusters.
    Generally speaking, the agglomeration phase is not very well understood so
far. The problem lies in the fact that molecular clusters of nanometer size are
difficult to measure. For smaller clusters one can use, e.g., mass spectrometry (see
above), for larger clusters of tens of nanometers light scattering techniques are
available. However, for the size range of interest here, diagnostic approaches are

missing so far. In addition, this growth phase is very fast which makes time resolved measurements difficult.

Thus, one is now in the need for a model to explain that the nanometer particles stick together to form larger clusters. The rapid time scale of agglomeration makes chemical reactions for this process unlikely. Models, see e.g. [14], suggest that the agglomeration is between small clusters of about 1 nm in size and larger, negatively charged clusters.

A large cluster of 20 nm diameter, say, should acquire a negative charge of about $Z_d \approx 30$ according to our rule of thumb Eq. (2.22). Even if electron depletion effects are considered, these particles will carry some negative charges.

For the small clusters of 1 nm diameter the situation is different. Measurements demonstrate that in the starting phase of the agglomeration the cluster density is very high (more than $10^9\,\mathrm{cm}^{-3}$ in Fig. 11.1). In comparison, the plasma density in these discharges typically is of the order of 10^8 to $10^9\,\mathrm{cm}^{-3}$, thus there are much more clusters (dust particles) than electrons. The electrons are dramatically depleted (see Sect. 2.6). The average charge on the dust particles is therefore very small, about 0.1 elementary charges.

Hence, for these small molecular clusters stochastic charge fluctuations become important, see Sect. 2.5.2. In that case, the particles are slightly negatively charged on average, but at certain times they can become neutral or even positive due to the random collection of electrons and ions. Naturally, the presence of small positively and large negatively charged particles at the same time in the plasma strongly enhances the agglomeration due to their Coulomb attraction. However, models have shown that even small *neutral* clusters agglomerate nearly as fast with larger negatively charged particles due the appearance of attractive image charges in the neutral particles [14].

Hence, as long as there are small clusters of 1 nm size present in the plasma they will agglomerate with the larger clusters. Eventually the clusters reach a size of a few tens nanometer in diameter, their number density drops to 10^7 or $10^8\,\mathrm{cm}^{-3}$. The particles acquire considerable net negative charges and due to their mutual Coulomb repulsion the particles cannot grow further by agglomeration.

11.1.3 Accretion Phase

After the rapid growth in the agglomeration phase it is found that the particles still grow further in size (see Fig. 11.1). The growth rate in this phase is often found to be close to the growth rate for the corresponding thin-film deposition on massive surfaces.

At this stage, the particles have typical sizes of tens of nanometers. They are negatively charged and trapped in the plasma bulk (see Sect. 3.7). Thus, effectively, the particles behave like small substrates on which additional layers of silicon-hydrogen films are deposited.

The growth of these particles can be continued until the particles reach the sub-micron or micrometer range. When the particles have grown to several micrometer size the particles experience stronger forces and may eventually drop out of the discharge, see Chap. 3.

11.2 Technological Impacts of Dusty Plasmas

Dusty plasmas can have "good" and "bad" implications on technology. Surface processing technologies are obviously concerned with particle-induced failures and it was recognized only until the end of the 1980s that (dusty) plasma processes could be involved in such dust contamination problems. Nowadays, a number of useful applications have emerged from dusty plasmas. Here, we like to illustrate some examples of technological applications of dusty plasmas and we will start with the problem of surface contamination by dust.

11.2.1 Surface Contamination by Dust

Plasma processes are widely used in technological applications such as plasma etching and sputtering or thin-film deposition. Numerous manufacturing steps for computer chip production include plasma processes. Although the plasma conditions are chosen in such a way that dust particle formation is unlikely they are optimized for highest growth rates of the film to be deposited. They are thus close to conditions where particle growth can occur. In these processing plasmas particles can grow either due to plasma polymerization as described above or particles etched or sputtered from the substrate can arrive in the plasma and can be trapped there (see Fig. 1.2 in the Introduction). When such dust particles are deposited onto the substrate during the discharge cycle or during switch off, these particles can do severe damages to the surface properties and can easily destroy integrated circuits. They can cause short circuits or may clog up wafer trenches (see Fig. 11.4 and [15, 16]).

Roughly speaking, such so-called "killer particles" are dangerous when they have about one third to one half the structure size. Current technology works with 300 mm wafers with structure sizes of 10 nm (2017), projected to 5 nm in 2020. Thus, presently, particles of 5 nm and larger pose a serious problem. As mentioned earlier, particles of that size cannot be detected by light scattering. This causes a severe diagnostic problem in the future when even smaller structures will be fabricated.

Current technologies to prevent particles from falling onto the substrates make use of the forces that have been discussed in Chap. 3. One possibility is to drill circular grooves in the electrode around the wafer. The grooves then finally lead to the vacuum port. The grooves disturb the sheath potential in such a way that the particles are preferably trapped above the groove and not above the wafer. The

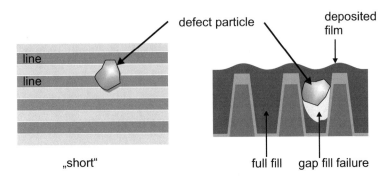

Fig. 11.4 Scheme of particle-induced failures on wafer structures: particle causing a short circuit between two signal lines and particle causing gap fill failure between etched trenches or ridges

grooves duct the dust particles to the vacuum port where the dust is sucked into the vent line.

Additional techniques directly use a gas puff that is blown across the wafer prior to plasma switch-off. The dust particles are blown into regions where they can do no harm to the wafer and when the plasma is switched off the particles settle onto the walls or outer regions of the electrode. For permanent removal of dust by gas flow one of the electrodes has many small holes through which the gas enters the discharge volume, a so-called showerhead configuration. On the other side of the chamber the gas is constantly evacuated. Thus, the discharge works with a constant gas flow through the chamber (most technological applications use discharges with permanent gas flows to remove sputter products and keep the discharge under constant conditions on the molecular level). The gas flow in the chamber is then directed in such way that possible dust particle are constantly washed out of the discharge.

Another way is to heat different parts of the chamber. The thermophoretic force drives the particles to colder plasma regions. Thus by heating the regions where no dust is wanted the particles are removed.

Which technique is applied in a specific discharge strongly depends on the exact conditions and a general statement is not possible here. But one should keep in mind that gas drag or thermophoretic forces scale with the particle radius as a^2, the electric force proportional to a. Thus, the above described techniques might not work efficiently when smaller and smaller dust particles have to be removed.

11.2.2 Dust in Fusion Devices

Dust in fusion devices occurs due to the interaction of the plasma with the plasma facing components (like graphite or carbon fibre composite (CFC) tiles) of the vessel. Dust can be produced by thermal overload of the surfaces leading e.g. to

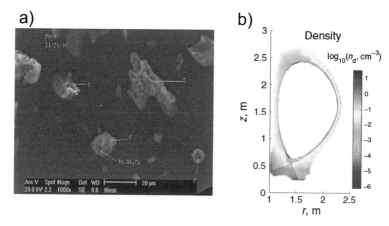

Fig. 11.5 (**a**) REM micrographs of dust collected from fusion devices. (**b**) Sketch of the poloidal cross section of a tokamak together with the simulated dust density. Reprinted figures from [19]. © (2010) John Wiley & Sons, Inc

brittle destruction of carbon, to melt layer loss of metals or to desintegration of codeposited layers, see e.g. [17–19]. Alternatively, dust can grow from sputtered carbon as in the low-temperature plasmas described above. Examples of dust particles collected in existing devices are shown in Fig. 11.5. Particles in active plasmas have also been seen from laser scattering experiments and with fast cameras.

The problems that are associated with dust in fusion devices are the following. First, dust can lead to difficulties with the vacuum vessel and pumping. Second, diagnostics might be covered by dust and cleaning inside the vacuum chamber of fusion devises is difficult. For example, mirrors for optical diagnostics of the plasmas might be "blinded" by the dust. More important, material eroded from the wall could enter the core plasma, e.g., in some tokamaks tungsten is used as a plasma facing component. As a high-Z material (the atomic number of tungsten is $Z = 74$), tungsten, when eroded from the wall into the core plasma, leads to strong bremsstrahlung losses and thus unwanted cooling of the plasma. Finally, in a real deuterium-tritium fusion plasma, the radioactive tritium can be chemically bound to carbonaceous dust particles (In the edge plasma the plasma conditions are not too different from those of low-temperature plasmas). The dust therefore adds to the radioactive inventory of the fusion device. For an existing tokamak, the density of carbonaceous dust in the edge plasma has been calculated from simulations [19, 20]. There it has been found highest in the divertor regions where the plasma is relatively cold. These issues are not problematic for existing devices. The potential implications for safety and operational performance in fusion devices such as ITER are under investigation [18].

11.2.3 Formation of New Materials

Plasmas in reactive, particle-forming gases are well suited to form new materials. On the one hand, as we have seen, particles can be easily trapped inside the plasma and their positions can be controlled by the various forces acting on them. On the other hand, the chemically active species allow to grow particles or to modify their surface properties.

When using plasmas to modify the surface properties of particles or to grow thin films on substrates one exploits the drastically different temperatures of the various species. The electrons are usually the hottest species with a temperature of a few eV. They are thus able to excite or ionize atoms and molecules, or to dissociate molecules into chemically very active species. Thus, the various radicals or other chemical active species are generated by the electrons. Thus, plasmas provide similar chemically active species as in hot wet-chemical reactants.

The ions and the neutral gas are typically at room temperature or at slightly elevated temperatures. In low-pressure plasmas the species with the highest density is by far the neutral gas. Thus, also the substrate and the particles are kept at low temperatures which is essential for the (thermal) stability of the substrate or the particles.

Consequently, in such reactive discharges one can have the advantage of high-temperature chemistry (due to the electrons) without the thermal stress on the particles or the substrates. This opens up the road to a large variety of new materials. However, in general, these processes are restricted to high value products and specific processes due to the small total mass of particles that can be treated.

Some examples of technological applications are listed below, see also [3, 21]:

- Dusty plasmas with nanometric particles at high particle density can be used in the deposition of silicon films (see also Sect. 11.2.4). The particles are deposited on the surface and lead to a significant increase in the deposition rate of thin silicon films without degradation in quality.
- Particles are coated with catalytic material. Due to the very large surface area of the particles, such systems provide very efficient catalysts. As an example, SiC particle are produced to about 80 nm in size by growth processes similar to the ones described above in Sect. 11.1. Then the reactive gas supply is stopped, but the particles remain trapped in the discharge. In a second step, palladium (Pd) is sputtered onto the trapped particles. For that purpose, a Pd wire in the discharge is negatively charged. By ion bombardment this wire serves as the metal source for the sputtering process. This finally leads to particles covered with Pd patches.
- Micron-sized iron particles are immersed into a plasma and trapped in the sheath of the discharge. Subsequently aluminum layers of about 100 nm thickness are sputtered onto the trapped iron particles using a magnetron source. Thereby the optical properties of the particles are changed, but they remain ferromagnetic and, hence, can be influenced by magnetic fields.

- Particles surfaces can be modified so that medical and pharmaceutical agents can attach to the surface of the particles. Again, the very large surface area of the particles leads to an efficient and controlled way to apply the medical drugs.
- Fluorescent particles are coated with a thin layers that keep the particles stable against bombardment from plasma particles. Such fluorescent particles can then be used in as the fluorescent layer in light tubes.

11.2.4 Polymorphous Silicon Films

As a final example for the technical potential of dusty plasmas, the deposition of thin silicon films for solar cells is presented. Plasma-deposited thin-film solar cells are often installed in (low-cost) consumer electronics such as pocket calculators or watches. These films are deposited in silane discharges on appropriate substrates under conditions where dust particle formation does not occur. This thin silicon film is responsible for the opto-electric properties of the solar cell. The film usually is amorphous and contains a lot of hydrogen. These films are denoted as a-Si:H, where the "a" stands for amorphous. The efficiency of such films is of the order of around 10%, only.[2]

Here, an interesting new development has emerged in the last recent years [22–25]. There, solar cells have been produced in silane discharges just like those that are used to deposit a-Si:H films. However, the discharges are operated under plasma conditions (gas pressure, gas flow, discharge power, substrate temperature etc.) that favor particle formation as discussed above. Previously, such conditions have been avoided since the formation of dust at thin film deposition was considered as very unfavorable.

Operating the discharge close to powder formation means that nevertheless particles of less than 10 nanometers in size are formed. Light-scattering techniques will characterize the discharge as dust-free since particles below 10 nanometers cannot be detected. Indirect methods, such as changes in the plasma impedance, can be used to infer particle production. These particles of less than 10 nm are then incorporated into the thin film at the substrate before the particles start to further grow in the agglomeration and accretion phase [22–25].

Such nanometer clusters are indeed incorporated in the film as can be seen in the micrograph in Fig. 11.6. The dust particles of nanometer size are crystalline and in the film they appear as tiny crystalline patches in the otherwise amorphous silicon matrix. The circles mark those crystalline regions. The dust particles serve as a crystalline nucleus. The resulting film is neither crystalline nor amorphous and consists of some crystalline patches in a still amorphous matrix. The film is still

---

[2]Efficiencies for commercially available multicrystalline Si solar cells for power production (on the roof of houses or in large solar cell fields) are around 14–19%. This is only achieved by crystalline silicon panels which, however, are relatively expensive.

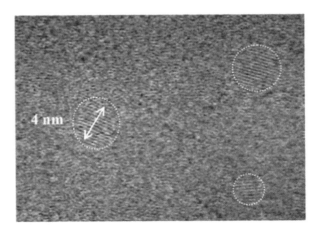

Fig. 11.6 Electron micrograph of a polymorphous silicon film at 150 °C. Reprinted from [22], ©
(2002), with permission from Elsevier

hydrogen-rich and is then termed as polymorphous silicon (pm-Si:H). So, these new
type of silicon film is due to the incorporation of nanometer-sized dust.

So, what are the properties of the pm-Si films? These polymorphous films have
superior hole transport properties and lower defect densities than standard a-Si:H.
This would suggest that solar cells from pm-Si:H show a better performance than
those from a-Si:H. However, it is found that pm-Si:H develops a very defective
interface between the positively-doped semiconductor material and the intrinsic pm-
Si:H layer, probably due to the bombardment by the charged silicon nanocrystals
and due to the high flux of hydrogen required to develop the film [25]. So, the overall
performance of pm-Si:H as solar cell material does not exceed that of standard a-
Si:H. However, recent modeling approaches demonstrate that pm-Si:H solar cell
with efficiencies of about 12% should be possible by reducing the defects at the
interface layer [25].

So with small changes to the manufacturing process from a-Si:H to dust-
containing pm-Si:H a serious improvement in the efficiency of solar cells might
be achievable using dusty plasmas.

References

1. A. Bouchoule (ed.), *Dusty Plasmas* (John Wiley & Sons, Chichester, 1999)
2. C. Hollenstein, Plasma Phys. Controlled Fusion **42**(10), R93 (2000)
3. L. Boufendi, A. Bouchoule, Plasma Sources Sci. Technol. **11**(3A), A211 (2002). http://stacks.
   iop.org/0963-0252/11/i=3A/a=332
4. L. Boufendi, M.C. Jouanny, E. Kovacevic, J. Berndt, M. Mikikian, J. Phys. D. Appl. Phys.
   44(17), 174035 (2011)

5. U. Kortshagen, Plasma Chem. Plasma Process. **36**, 73 (2016). https://doi.org/10.1007/s11090-015-9663-4
6. C. Hollenstein, J.L. Dorier, J. Dutta, L. Sansonnens, A.A. Howling, Plasma Sources Sci. Technol. **3**(3), 278 (1994). http://stacks.iop.org/0963-0252/3/i=3/a=007
7. C. Courteille, C. Hollenstein, J.L. Dorier, P. Gay, W. Schwarzenbach, A.A. Howling, E. Bertran, G. Viera, R. Martins, A. Macarico, J. Appl. Phys. **80**, 2069 (1996)
8. A. Bouchoule, L. Boufendi, Plasma Sources Sci. Technol. **2**, 204 (1993). http://stacks.iop.org/0963-0252/2/i=3/a=011
9. A.A. Howling, L. Sansonnens, J.L. Dorier, C. Hollenstein, J. Appl. Phys. **75**(3), 1340 (1994). https://doi.org/10.1063/1.356413
10. C. Hollenstein, W. Schwarzenbach, A.A. Howling, C. Courteille, J.L. Dorier, L. Sansonnens, J. Vac. Sci. Technol. A **14**(2), 535 (1996). https://doi.org/10.1116/1.580140
11. C. Hollenstein, A. Howling, C. Courteille, D. Magni, S.M. Scholz, G. Kroesen, N. Simons, W. de Zeeuw, W. Schwarzenbach, J. Phys. D: Appl. Phys. **31**, 74 (1998)
12. A.J. Adamczyk, M.F. Reyniers, G.B. Marin, L.J. Broadbelt, Theor. Chem. Acc. **128**, 91 (2011). https://doi.org/10.1007/s00214-010-0767-x
13. J. Sobol-Antosiak, W. Ptak, Mater. Lett. **56**, 842 (2002). https://doi.org/10.1016/S0167-577X(02)00625-0. http://www.sciencedirect.com/science/article/pii/S0167577X02006250
14. L. Ravi, S.L. Girshick, Phys. Rev. E **79**, 026408 (2009). https://doi.org/10.1103/PhysRevE.79.026408. https://link.aps.org/doi/10.1103/PhysRevE.79.026408
15. D.W. Cooper, Aerosp. Sci. Technol. **5**(3), 287 (1986). https://doi.org/10.1080/02786828608959094
16. S.W. Jones, *Introduction to Integrated Circuit Technology: Fifth Edition* (IC Knowledge LLC, Georgetown, 2001)
17. J. Winter, G. Gebauer, J. Nucl. Mater. **266-269**, 228 (1999)
18. J. Sharpe, D. Petti, H.W. Bartels, Fusion Eng. Des. **63-64**(0), 153 (2002)
19. S.I. Krasheninnikov, A.Y. Pigarov, R.D. Smirnov, T.K. Soboleva, Contrib. Plasma Phys. **50**(3–5), 410 (2010)
20. R.D. Smirnov, A.Y. Pigarov, M. Rosenberg, S.I. Krasheninnikov, D.A. Mendis, Plasma Phys. Controlled Fusion **49**(4), 347 (2007). http://stacks.iop.org/0741-3335/49/i=4/a=001
21. H. Kersten, R. Wiese, G. Thieme, M. Fröhlich, A. Kopitov, D. Bojic, F. Scholze, H. Neumann, M. Quaas, H. Wulff, R. Hippler, New J. Phys. **5**, 93 (2003)
22. P.R. i Cabarrocas, A.F. i Morral, Y. Poissant, Thin Solid Films **403–404**, 39 (2002). https://doi.org/10.1016/S0040-6090(01)01656-X, http://www.sciencedirect.com/science/article/pii/S004060900101656X. Proceedings of Symposium P on Thin Film Materials for Photovoltaics
23. P.R. i Cabarrocas, N. Chabane, A.V. Kharchenko, S. Tchakarov, Plasma Phys. Controlled Fusion **46**(12B), B235 (2004). http://stacks.iop.org/0741-3335/46/i=12B/a=020
24. Y. Poissant, P. Chatterjee, P. Roca i Cabarrocas, J. Appl. Phys. **94**, 7305 (2003). https://doi.org/10.1063/1.1623610
25. Abolmasov, Sergey, Cabarrocas, Pere Roca i, Chatterjee, Parsathi, EPJ Photovolt. **7**, 70302 (2016). https://doi.org/10.1051/epjpv/2015011

Chapter 12
Astrophysical Dusty Plasmas

Historically, the field of dusty plasma physics has originated from the investigations of dusty plasmas in astrophysical situations, from the study of comets, interstellar clouds or the rings of the Giant Planets [1, 2].

In space, dust particles can collect electric charges by various processes, as discussed in Chap. 2, like electron and ion collection, secondary emission, photoelectric charging etc. The particles react to the magnetic and electric fields in the plasma environment of planets, the solar wind and gravitational forces. The combination of these processes give rise to a description of dust processes in astrophysical situations that is sometimes referred to as "gravito-electrodynamics".

Space plasmas are almost always dusty plasmas. In fact, it is difficult to find astrophysical plasmas that do not include dust particles (maybe except the fusion plasmas in the interior of stars). There is a vast amount of possibilities where astrophysical and extraterrestrial dusty plasmas naturally occur.

When starting at Earth we find the so-called noctilucent clouds and polar mesospheric summer echoes (PMSE) in the ionosphere. In our solar system we find dust near the surface of the Moon [3, 4], e.g. as "horizon glow", but also as a permanent dust cloud around the Moon at altitudes of about 200 km [5]. Further examples are dust in cometary tails, like the famous comet 67P/Churyumov–Gerasimenko studied with the Rosetta mission [6], and dust forming the zodiacal light. Also, interstellar dust-containing hydrogen or molecular clouds also add to this list.

From the vast variety of astrophysical dust systems we will restrict here to dust near the Giant planets Jupiter and Saturn, especially to the formation of spokes in the rings of Saturn and dust particle "orbits" at Jupiter and Saturn. These are situations where experimental data are available from the spacecraft Galileo at Jupiter or Cassini at Saturn. However, first, a few words on the noctilucent clouds will be given here. Further aspects of space dusty plasmas are treated in [1, 2, 7–10].

© Springer Nature Switzerland AG 2019
A. Melzer, *Physics of Dusty Plasmas*, Lecture Notes in Physics 962,
https://doi.org/10.1007/978-3-030-20260-6_12

12.1 Noctilucent Clouds and PMSE

Noctilucent clouds are "nightly shining" clouds that are observable in the summer months (May to September) at high-latitude regions (50–80°) of the Earth, like Northern Europe [11, 12]. They can be seen long after sunset. They are thus located at high altitudes of 80–90 km above the surface of the Earth where the sun, although being already below the horizon, still illuminates the clouds (see Fig. 12.1a).

The clouds are thus located in the so-called mesosphere or the lower edge of the ionosphere. They are assumed to consist of water (or methane) ice particles which are charged by the plasma of the ionosphere. The particle size is below 100 nm. Surprisingly, in the summer months the atmosphere at these altitudes is extremely cold (<140 K) and is much colder than during winter times. This low temperature allows water (and methane) to freeze under the conditions of the upper atmosphere preferably on small condensation nuclei. This explains why the clouds are preferably visible in the summer months.

Noctilucent clouds have first been reported in 1885, a few years after the big explosion of the Krakatoa volcano in 1883. It is not clear whether noctilucent clouds

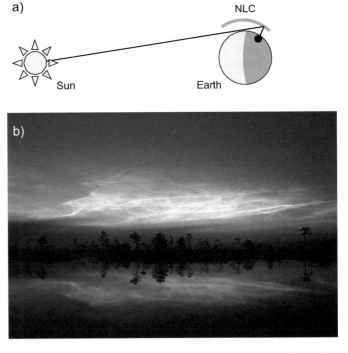

Fig. 12.1 (**a**) Scheme of the observation geometry. The observer on the night side of the Earth (marked by the dot) can see the noctilucent cloud illuminated by the sun. (**b**) Photo of a noctilucent cloud over Kuresoo bog, Viljandimaa, Estonia by Martin Koitmäe, distributed under GFDL or CC BY-SA 4.0, from Wikimedia Commons

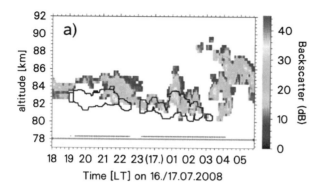

Fig. 12.2 Simultaneous occurrence of PMSE and noctilucent clouds over Northern Norway. Color-coded is the radar backscatter signal as a function of time and height. The contour lines denote the noctilucent cloud position determined from the LIDAR. From [16] distributed under the Creative Commons Attribution 3.0 License

have been present before this explosion, but have not been reported, or whether the Krakatoa has triggered the existence of the noctilucent clouds by blowing large amounts of dust into the atmosphere. Nevertheless, the occurrence and strength of noctilucent clouds have increased over the years since their discovery. It is sometimes argued that the increased methane release on Earth leads to the increased occurrence of noctilucent clouds [13].

The "dust" particles in the noctilucent clouds can be observed by sounding rockets launched into the ionosphere, by lasers fired into the atmosphere, so-called LIDARs (LIght Detection and Ranging) and by radar (see Fig. 12.2 and [14, 15]).

With LIDAR technique the laser light is back-scattered from particles in the atmosphere and is detected at different places on the Earth surface. The height of the scattering particles is found from triangulation. LIDARs are sensitive to dust particles larger than approximately 30 nm due to the use of scattering of (mainly) visible light (see Chap. 10). Hence, LIDARs map out the noctilucent clouds.

Similarly, using radar bursts sent into the atmosphere, the back-scattered radar signal is observed. From the run-time difference between sending and receiving the radar signals the height of the backscattering agents is derived. Since radars operate in the frequency range of typically 10 MHz to 1 GHz (with corresponding wavelengths of 0.3 m to 30 m), radars do not "see" the dust particles directly, but the radar signal is possibly scattered from the electron and ion clouds around the dust particles or from turbulent plasma structures [15]. Radar backscatter usually is an indication of smaller dust particles of 1–20 nm size. Often during summer months one observes radar backscatter in the mesosphere (up to 100 km altitude), the above mentioned PMSEs. It is reasonable to assume that the PSMEs are related to the presence of noctilucent clouds and vice versa. Indeed, there are numerous observations that substantiate this reasoning, see Fig. 12.2. However, one also finds noctilucent clouds and no radar backscatter, or the other way round. This might be due to the fact that the presence of PMSEs and noctilucent clouds rely on different particle sizes, larger than about 30 nm for the noctilucent clouds and smaller than about 30 nm for radar backscatter. The physics behind the noctilucent clouds and PMSEs is, however, not fully understood so far [12].

12.2   Dust Streams from Jupiter

We now turn to examples of dusty plasma appearances in our solar system and will discuss dust properties near Jupiter and Saturn [17].

Jupiter is the largest planet of our solar system (its mean radius is $R_J = 69,911$ km) and the one with the highest mass ($m_J = 1.898 \times 10^{27}$ kg). One would therefore expect that due to gravitational attraction all dust in its vicinity would be "sucked" towards Jupiter. Nevertheless, the Ulysses spacecraft has discovered a high-speed dust flux away from the Jovian system in 1992. This dust stream has been investigated in much more detail by the Galileo spacecraft since 1996. Galileo had a dust detector on board that was able to measure the velocity and the mass of the arriving dust particles and was calibrated for dust particles in the size range 30 nm $< a <$ 100 nm. The dust detector also measured a certain fraction of quite large particles (>1 μm). However, the majority of the particles were very small (smaller than the calibration range, i.e. $a < 30$ nm) and very fast. These particles were generally directed away from Jupiter [17, 18].

On its path, Galileo measured the dust particles arriving at the dust detector (see Fig. 12.3). Very high impact rates were found close to the moon Io. Io is an active volcanic moon and is closest to Jupiter. So, Io was already very early considered to be the source of the dust particles emerging from the Jovian system. Dust from the volcanic eruptions is sent into the magnetosphere of Jupiter and is transported outward.

To describe this outward motion the dust particle dynamics has to be modeled quite accurately. The equation of motion of a dust particle in the vicinity of Jupiter can be modeled by [17]

$$\ddot{\mathbf{r}} = \frac{Q_d}{m_d} \left(\dot{\mathbf{r}} \times \mathbf{B} + \mathbf{E}_c \right) - G m_J \frac{1}{r^3} \mathbf{r}, \qquad (12.1)$$

where G is the gravitational constant. Here, the electric field force, the Lorentz force and gravitational forces have to be considered. The magnetosphere would rigidly co-rotate with the planet at the rotation frequency Ω_J if the magnetosphere was perfectly conducting. Further, $\mathbf{B}$ is the magnetic field of Jupiter and $\mathbf{E}_c = (\mathbf{r} \times \Omega_J) \times \mathbf{B}$ is the co-rotational electric field. Other forces can be neglected, here.

As the grain moves through the plasma the particle charge is determined by the different currents to the particle, see Eq. (2.1),

$$\frac{dQ_d}{dt} = \sum_k I_k$$

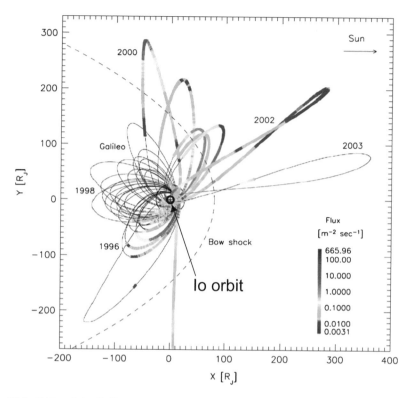

Fig. 12.3 Orbits of the Galileo spacecraft from 1996 to 2003 projected onto Jupiter's equatorial plane. Superimposed on the path are the fluxes of Jovian dust stream particles. High impact rates are found near the moon Io. Reprinted from [19] with permission from Elsevier

where typically electron and ion collection, secondary emission and photoelectron emission are considered. The particle charge is in local equilibrium, but the charging currents depend on the local plasma parameters. The plasma parameters and the magnetic field **B** (and thus the co-rotational electric field $\mathbf{E_c}$) around Jupiter are quite accurately known and are used to calculate the charges and the Lorentz force on the dust particles. It is important to note, here, that the magnetic field axis and the rotational axis of Jupiter have a relative angle of about 10°. Thus during Jupiter's rotation the magnetic field precesses, i.e. has a "tumbling" motion.

The equation of motion for the dust particles is then solved numerically. Dust particles of various sizes are considered to originate from Io's orbit and their motion through the Jovian magnetosphere is followed in the simulation [17, 20].

The trajectories of particles in the size between 5 and 100 nm emerging from the orbit of the moon Io have been calculated [20], see Fig. 12.4a, b. One can see that indeed particles are emitted outwards away from Jupiter, preferably particles in the size range between 5 and 15 nm. These particles leave Io and form a comet-tail like swarm that spirals outward. One can also see that the particles leave the

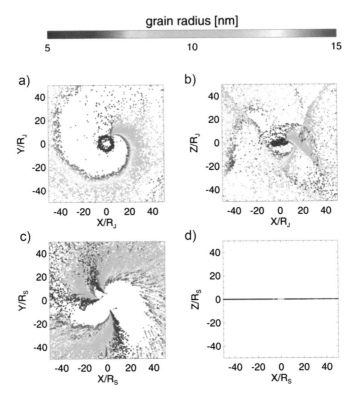

Fig. 12.4 Modeled positions of dust grains ejected from Io at Jupiter projected onto the equatorial plane (**a**) and the meridional plane (**b**). The color-scale corresponds to particle size in the range between 5 and 15 nm. (**c, d**) Same as (**a, b**) for dust ejected from the moons Dione, Helene, and Rhea at Saturn. Reprinted figure from [21]. © (2004) John Wiley & Sons, Inc

equatorial plane and acquire positions quite far above and below the equatorial plane, see Fig. 12.4b. This is due to the inclination of the magnetic field axis relative to Jupiter's rotation axis. So, the dust particles form a pattern around Jupiter like a ballerina skirt. The particles on their way outward usually have positive charges due to photoelectron emission, the particle potential was found to be around $\phi_p \approx +5$ V.

12.3   Dust Orbits at Saturn

The situation of dust particle dynamics at Saturn is similar to that at Jupiter. Therefore, similar dust streams are also found by the dust detector on the Cassini spacecraft that orbited around Saturn from 2004 to 2017 [22]. From the models [20], the source of the dust particles can be the moons Dione (plus Helene) and Rhea which orbit in the outskirts of the magnetosphere of Saturn (their distances

Fig. 12.5 "Halo" orbits of dust particles above the equatorial plane at Saturn. Shown is the action of the three responsible forces: gravity, centrifugal force and electromagnetic forces. After [21]

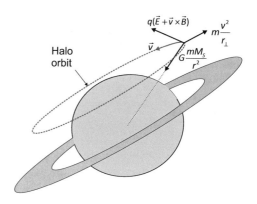

from Saturn are $6.5R_S$ and $9.0R_S$ with $R_S = 58,232$ km being the radius of Saturn, $m_S = 568 \times 10^{24}$ kg). These moons are not volcanic, but dust may enter the Saturnian magnetosphere by meteroid bombardment. Recently, is has been found that the moon Enceladus at $4.0R_S$ emits a plume of gas and dust that feeds Saturn's E-ring [23] and might contribute to the dust streams.

The simulated dust transport for Saturn is shown in Fig. 12.4c, d. In contrast to dust at Jupiter, in Saturn's system the particle motion is purely radially outward where the particles stay in the equatorial plane. This is due to the fact that the magnetic field axis of Saturn is closely aligned with its rotation axis.

The aligned magnetic field also gives rise to a new phenomenon that is not observable at Jupiter. Since the situation with aligned magnetic field and rotation axis at Saturn is very symmetric, stable particle orbits might exist that are high above the equatorial plane and that never cross this plane, so-called "halo" orbits, see Fig. 12.5. Particles in these orbits are kept at such positions due to the balance of three forces: the centrally inward gravitational force, the radially outward centrifugal force and the electromagnetic force that points upward under a certain angle (see Fig. 12.5).

12.4   Spokes in Saturn's Rings

Next, the formation of spokes will be discussed. We have mentioned this example as a paradigm for dusty plasma physics in astrophysical situations in the Introduction (Chap. 1). And, indeed, the problem of spoke formation can only be treated when the various processes in dusty plasmas, like charging, forces etc., are taken into account.

It is interesting to note that different theories exist to explain the formation of spokes and a definite answer cannot be given at the moment. It is expected that the recent mission of spacecraft Cassini will allow a deeper insight into the spoke formation.

Spokes are known since the early 1980s from photos of the spacecrafts Voyager 1 and 2 and they have been rediscovered by Cassini in 2005[1] [24, 25]. The spokes appear in the B-ring of Saturn which extends from about 92,000 km to 120,000 km and is the most opaque ("dense") and most prominent ring of Saturn. Spokes are radial features in the B-ring, see Fig. 1.1a). The spokes appear dark when viewed in backscattered light and bright in forward scattered light. This suggests that the spokes consist of sub-micron particles which have such scattering properties. Spokes come into existence in less than 5 min and disappear after about 5 h. The radial elongation of the spokes is a few 10^4 km, their width is between 200 and 1000 km. The spokes are therefore a very dynamical phenomena and are unlikely to be explained by gravitational effects, only.

To illustrate the various dusty plasma aspects that might come into play here, the model of Goertz and Morfill [26] will be discussed as one possible model to explain spoke formation, see also [21]. In this model it is assumed that spokes become visible when dust particles are lifted above ring plane. In the ring the small dust particles are not visible due to the larger rocks and boulders (of 10 cm to 10 m size). Above the ring plane the particles become visible and show the scattering properties as described above. To be lifted above the ring plane the dust particles must be accelerated. In this model the acceleration is due to a perpendicular (vertical) electric field. Now the question arises how such an electric field appears and why its appearance is only sporadic (spokes do not appear always and everywhere).

Under normal conditions the plasma density in the ring plane is small ($n_e \leq 10^4$ cm^{-3}) and the surface potential of the dust is slightly positive on the sun-facing side, the mean charge of sub-micron dust is expected to be much smaller than one elementary charge. Hence, the dust is essentially neutral. In the shadow of Saturn, the dust potential is expected to be around -6 V since there is no photoemission. The ring plane is an equipotential line. Due to the small dust charge and the small electric fields at the ring surface dust particles cannot be lifted above the ring plane.

It is then assumed that a local increase of plasma density can change this equilibrium situation drastically. Such a local plasma density increase can be due to meteoretic impacts into the ring plane. In the plasma with higher density the surface potential of the dust becomes negative (around -6 V). Furthermore, the equipotential lines become compressed below the local plasma (see Fig. 12.6). This can accelerate dust particles above the ring plane where they become visible.

Having now elevated dust particles there is the need to explain the radial motion of the spokes. There, one has to look into the rotation of the different species. The higher-density plasma cloud is coupled to the magnetosphere of Saturn. Since the entire magnetosphere of Saturn rotates at the same frequency Ω_S as Saturn also this plasma cloud will rotate at this speed. The dust particles however will rotate around Saturn on Kepler orbits with the Keplerian frequency $\Omega_K > \Omega_S$.[2] Thus, the plasma

---

[1]Interestingly, the spokes were not visible when Cassini arrived at Saturn in early 2004.

[2]The particles in the B-ring rotate faster than Saturn. Saturn performs one revolution in 10.2 h, the Kepler period at a distance of 100,000 km from Saturn is about 9 h.

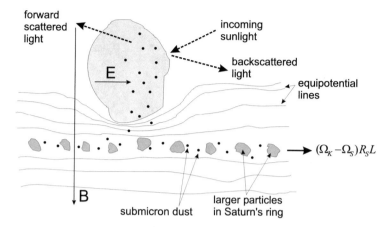

Fig. 12.6 Model for the levitation of dust above the ring plane and the formation of spokes. Under a locally increased plasma density (dark grey) the equipotential lines are compressed leading to an increased electric field. The particles are lifted above the ring plane. There, the particles move on Kepler orbits and accumulate on one side of the local plasma disturbance and are then transported radially (see text). The view in this Figure is radially away from Saturn. After [21, 26]

cloud and the particles will rotate at different speeds, so the dust is accumulated on one side of the plasma cloud (see Fig. 12.6). The particles cannot leave the plasma cloud due to confining electric fields at the plasma cloud boundary.

Since the plasma cloud is quasineutral and the negative particles accumulate on one side a secondary electric field E arises. This electric field together with the magnetic field of Saturn (which is nearly vertical near the ring plane) leads to an $E \times B$ drift in the radial direction. Thus, the plasma cloud with the particles will stretch radially and radially elongated dust structures are formed, which is seen as the spokes.

The particles lifted above the ring plane should have sizes in the range of 100 to 300 nm. Smaller particles need much more time to charge up (the charging time is indirect proportional to particle size, see Eq. 2.24). Larger particles have too large mass and are not accelerated fast enough. This size range is just the one expected from the scattering of the sunlight. Thus the model makes a number of predictions which are in agreement with the findings from the observations.

In a different model, Bliokh and Yaroshenko [27] explain the spoke formation by density waves in a multi-stream situation. This approach also allows to account for the very many narrow rings and gaps within the B-ring. Although such dust density waves are probably not strong enough to fully explain the spoke formation they represent an intriguing mechanism. This demonstrates that the question of spoke formation is not settled yet.

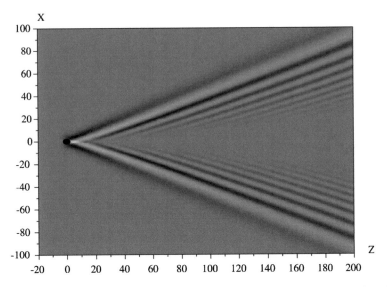

Fig. 12.7 Modeled Mach cone in Saturn's rings. Reprinted from [29], with the permission of AIP Publishing

12.5   Mach Cones in Saturn's Rings

Finally, we like to address an interesting diagnostic method that might be able to measure plasma properties in the rings of Saturn. This idea was first proposed by Havnes et al. in 1995 [28].

There, the effect of Mach cone formation in Saturn's rings is considered. As mentioned above, large boulders of centimeter to meter size move at Keplerian velocities around Saturn. At a radial distance of $R = 100,000$ km the Keplerian velocity is around 19 km/s. Small dust particles are coupled to the magnetosphere of Saturn that co-rotates with the planet.[3] At that distance this velocity is around 17 km/s. Thus there is a large relative velocity of larger boulders and small dust particles. This velocity difference might even be larger than the sound speed of dust-acoustic waves in the ring. Thus, Mach cones might emerge in Saturn's rings by launching dust-acoustic waves. Since the dust is expected to be only weakly coupled in Saturn's ring the dust-acoustic wave is the type of wave with the smallest sound speed and thus will be preferably excited by boulders that move with supersonic speed relative to the smaller dust particles. Simulated Mach cones of dust-acoustic waves in the rings of Saturn are shown in Fig. 12.7.

These Mach cones will be very similar to those observed in strongly coupled dust systems as described in Sect. 7.6 The main difference is that in strongly coupled systems we find Mach cones due to dust lattice wave excitation whereas in Saturn's

---

[3]This statement somewhat contradicts that in the previous section.

rings we expect Mach cones of dust-acoustic waves. As for the DLW Mach cone, the wave pattern of the DAW Mach cone shows a complicated structure with fine details due to the dispersion of the DAW and other effects like spatially and temporally varying dust densities or dust charges. From the Mach cones and its detailed structure the properties of the dusty plasma inside the rings can be derived. Again, it is anticipated that Mach cones might be observed from the data of the Cassini mission.

References

1. K.H. Glaßmeier, M. Scholer, *Plasmaphysik im Sonnensystem* (BI–Wissenschaftsverlag, Mannheim, Wien, Zürich, 1991)
2. C.K. Goertz, Rev. Geophys. **27**, 271 (1989)
3. E. Grün, M. Horányi, Z. Sternovsky, Planet. Space Sci. **59**, 1672 (2012)
4. S.I. Popel, S.I. Kopnin, A.P. Golub', G.G. Dol'nikov, A.V. Zakharov, L.M. Zelenyi, Y.N. Izvekova, Sol. Syst. Res. **47**, 419 (2013). https://doi.org/10.1134/S0038094613060063
5. M. Horányi, J.R. Szalay, S. Kempf, J. Schmidt, E. Grün, R. Strama, Z. Sternovsky, Nature **522**, 324 (2015). https://doi.org/10.1038/nature14479
6. G.H. Jones, M.M. Knight, A. Fitzsimmons, M.G.G.T. Taylor, Philos. Trans. R. Soc. A Math. Phys. Eng. Sci. **375** (2017). http://doi.org/10.1098/rsta.2017.0001
7. D.A. Mendis, M. Rosenberg, Annu. Rev. Astron. Astrophys. **32**, 419 (1994). https://doi.org/10.1146/annurev.aa.32.090194.002223
8. P. Bliokh, V. Sinitsin, V. Yaroshenko, *Dusty and Self-Gravitational Plasma in Space* (Kluwer Academic Publishers, Dordrecht, 1995)
9. I. Mann, N. Meyer-Vernet, A. Czechowski, Phys. Reports **536**, 1 (2014). https://doi.org/10.1016/j.physrep.2013.11.001
10. F. Verheest, *Waves in Dusty Space Plasmas* (Kluver Academic Publishers, Dordrecht, 2000)
11. M. Gadsden, W. Schröder, *Noctilucent Clouds* (Springer, Berlin, 1989). https://doi.org/10.1007/978-3-642-48626-5
12. O. Havnes, in *Dusty Plasmas in the New Millenium*, vol. 649, ed. by R. Bharuthram, M.A. Hellberg, P. Shukla, F. Verheest (AIP Conference Proceedings, Melville, New York, 2002), p. 13
13. F.J. Lübken, U. Berger, G. Baumgarten, Geophys. Res. Lett. **45**(13), 6681 (2018). https://doi.org/10.1029/2018GL077719
14. V. Nussbaumer, K.H. Fricke, M. Langer, W. Singer, U. von Zahn, J. Geophys. Res. **101**, 19161 (1996)
15. J.Y.N. Cho, J. Röttger, J. Geophys. Res.-Atmos. **102**, 2001 (1997). https://doi.org/10.1029/96JD02030
16. N. Kaifler, G. Baumgarten, J. Fiedler, R. Latteck, F.J. Lübken, M. Rapp, Atmos. Chem. Phys. **11**(4), 1355 (2011). https://doi.org/10.5194/acp-11-1355-2011
17. M. Horányi, AIP Conf. Proc. **649**(1), 22 (2002). https://doi.org/10.1063/1.1527733
18. H. Krüger, D. Bindschadler, S. Dermott, A. Graps, E. Grün, B. Gustafson, D. Hamilton, M. Hanner, M. Horányi, J. Kissel, D. Linkert, G. Linkert, I. Mann, J. McDonnell, R. Moissl, G. Morfill, C. Polanskey, M. Roy, G. Schwehm, R. Srama, Planet. Space Sci. **58**, 965 (2010). https://doi.org/10.1016/j.pss.2010.03.003
19. H. Krüger, G. Linkert, D. Linkert, R. Moissl, E. Grün, Planet. Space Sci. **53**, 1109 (2005). https://doi.org/10.1016/j.pss.2005.04.009
20. M. Horányi, Phys. Plasmas **7**, 3847 (2000). https://doi.org/10.1063/1.1288909.

21. M. Horányi, T.W. Hartquist, O. Havnes, D.A. Mendis, G.E. Morfill, Rev. Geophys. **42**(4), RG4002 (2004). https://doi.org/10.1029/2004RG000151
22. R. Srama, S. Kempf, G. Moragas-Klostermeyer, S. Helfert, T. Ahrens, N. Altobelli, S. Auer, U. Beckmann, J. Bradley, M. Burton, V. Dikarev, T. Economou, H. Fechtig, S. Green, M. Grande, O. Havnes, J. Hillier, M. Horányi, E. Igenbergs, E. Jessberger, T. Johnson, H. Krüger, G. Matt, N. McBride, A. Mocker, P. Lamy, D. Linkert, G. Linkert, F. Lura, J. McDonnell, D. Möhlmann, G. Morfill, F. Postberg, M. Roy, G. Schwehm, F. Spahn, J. Svestka, V. Tschernjawski, A. Tuzzolino, R. Wüsch, E. Grün, Planet. Space Sci. **54**, 967 (2006). https://doi.org/10.1016/j.pss.2006.05.021
23. F. Postberg, N. Khawaja, B. Abel, G. Choblet, C.R. Glein, M.S. Gudipati, B.L. Henderson, H.W. Hsu, S. Kempf, F. Klenner, G. Moragas-Klostermeyer, B. Magee, L. Nölle, M. Perry, R. Reviol, J. Schmidt, R. Srama, F. Stolz, G. Tobie, M. Trieloff, J.H. Waite, Nature **558**, 564 (2018). https://doi.org/10.1038/s41586-018-0246-4
24. B.A. Smith, L. Soderblom, R. Beebe, J. Boyce, G. Briggs, A. Bunker, S.A. Collins, C.J. Hansen, T.V. Johnson, J.L. Mitchell, R.J. Terrile, M. Carr, A.F. Cook, J. Cuzzi, J.B. Pollack, G.E. Danielson, A. Ingersoll, M.E. Davies, G.E. Hunt, H. Masursky, E. Shoemaker, D. Morrison, T. Owen, C. Sagan, J. Veverka, R. Strom, V.E. Suomi, Science **212**(4491), 163 (1981). https://doi.org/10.1126/science.212.4491.163
25. C.J. Mitchell, M. Horányi, O. Havnes, C.C. Porco, Science **311**(5767), 1587 (2006). https://doi.org/10.1126/science.1123783
26. C.K. Goertz, G. Morfill, Icarus **53**, 219 (1982)
27. P.V. Bliokh, V.V. Yaroshenko, Sov. Astron. **29**, 330 (1985)
28. O. Havnes, T. Aslaksen, T.W. Hartquist, F. Li, F. Melandsø, G.E. Morfill, T. Nitter, J. Geophys. Res. **100**, 1731 (1995)
29. A. Brattli, O. Havnes, F. Melandsø, Phys. Plasmas **9**, 958 (2002)

Chapter 13
Summary

In these Lecture notes an overview over the various effects in dusty plasmas has been given. Fundamental properties of dusty plasma like particle charging, interaction potentials and forces on the dust have been presented. Waves in weakly and strongly coupled dusty plasmas have been discussed and the extension towards normal modes in finite systems has been demonstrated. Also, an introduction into magnetic field effects in dusty plasmas has been given together with a collection of diagnostic techniques for dusty plasmas. Finally, a brief description of applications of dusty plasmas in technology has been presented, before some examples from astrophysical situations have been mentioned.

Dusty plasmas require the development of novel diagnostic techniques for the observation of the full phase-space dynamics of individual microparticles as well as the collective behavior of submicron and nanometer particles. The progress in imaging technology allows to enter new regimes of dusty plasmas and promotes advanced image analysis and data handling techniques.

To summarize, the main properties of dusty plasmas compared to "usual" plasmas are compiled below again:

- Dusty plasmas are at least three component plasmas (electrons, ions and dust). In this sense, dusty plasmas are somewhat comparable to negative ion plasmas.
- However, the typical charge on the charge carriers (dust) are of the order of 10,000 elementary charges which leads to strong coupling on the one hand and to strong reactions to electric fields on the other hand.
- The dust charge is variable and depends on the local plasma parameters. The charging time of particles is finite. Thus, the charge becomes a dynamic variable and can lead to novel dynamic phenomena.
- The dust mass is by orders of magnitudes larger than that of electrons and ions. Thus the dominant time scale is that of the dust plasma frequency ω_{pd} which is by orders of magnitude smaller than that of electrons and ions leading to convenient time scales for the observation of dynamic processes in laboratory discharges.

© Springer Nature Switzerland AG 2019
A. Melzer, *Physics of Dusty Plasmas*, Lecture Notes in Physics 962,
https://doi.org/10.1007/978-3-030-20260-6_13

- The separation of time and mass scales leads to new types of waves and dynamical phenomena.
- The slow dust time scales allow that electrons *and* ions contribute to shielding which should result in different shielding scales.
- The dust size is not negligibly small leading to surface phenomena and forces which are unimportant in "usual" plasmas.
- Due to the high dust mass magnetization of the dust component might be possible only under very strong fields using submicron dust particles. Hence, dust magnetization is intrinsically connected with magnetization of electrons and ions which is a relatively unexplored situation in low-temperature plasma physics.

Due to all of these unique properties of dusty plasmas a number of new phenomena occur in dusty plasmas like new force, new types of waves, crystallization processes, phase transitions, observation of processes on the kinetic level and many more. We hope that we have clarified the origin of these phenomena and that we have demonstrated why dusty plasmas have become one of the very interesting fields in plasma physics.

Glossary

Lists of Abbreviations

BCC	Body-centered cubic
CCD	Charge-coupled device, camera type
CMOS	Complementary metal-oxide-semiconductor, camera type
DAW	Dust-acoustic wave
DIAW	Dust ion-acoustic wave
DLW	Dust-lattice wave
FCC	Face-centered cubic
HCP	Hexagonal close-packed
HMDSO	Hexamethyldisiloxane
ISS	International Space Station
LIDAR	Light detection and ranging
OCP	One-component plasma
OML	Orbital motion limit
PIV	Particle image velocimetry
PMSE	Polar mesospheric summer echo
RF	Radio frequency

© Springer Nature Switzerland AG 2019
A. Melzer, *Physics of Dusty Plasmas*, Lecture Notes in Physics 962,
https://doi.org/10.1007/978-3-030-20260-6

Symbol	Description	See Section
a	Particle radius	
a	Acceleration	
A	Dynamical matrix	8.4
b	Impact parameter	2.2.1
b	Interparticle distance	5.1
b_{WS}	Wigner-Seitz radius	5.1
b_{c}	Impact parameter for charge collection	2.2
$b_{\pi/2}$	Impact parameter for 90° deflection	3.3
B	Magnetic field strength	
c	Sound speed	6.2, 6.5, 7.2
C	Capacitance of dust particle	2.4.2
C, C'	Camera	10.6.2
d	Vertical distance between two dust layers	4.2
d_{f}	Vertical distance to ion focus	4.3
e	Elementary charge	
E	Electric field strength	
E	Energy of dust clusters	8.3
E_0	Normalized energy of dust clusters	8.3
f	Focal length	10.6.1
F	Force	
F	Fundamental matrix	10.6.2
g	Gravitational acceleration	3.1
$g(r)$	Pair correlation function	5.5
$g_6(r)$	Bond angle correlation function	5.5
G	Gravitational constant	12.2
h_{e}	Electron Hall parameter	9.1
h_{i}	Ion Hall parameter	9.1
h_{d}	Dust Hall parameter	9.1
k	(Effective) spring constant	7.1, 5.5
k_{B}	Boltzmann's constant	
k_{n}	Thermal conductivity of neutral gas	3.5
K	Camera matrix	10.6.1
I_{e}	(OML) Electron charging current	2.2
I_{i}	(OML) Ion charging current	2.2
I_{S}	Secondary electron charging current	2.3.2
I_k	Image pixel intensity	10.2
ℓ_{mfp}	Mean free path	2.2.4
L	Damping length	6.2
n_{d}	Dust density	
n_{e}	Electron density	

(continued)

Symbol	Description	See Section
n_i	Ion density	
n_n	Neutral gas density	
m_d	Dust mass	
m_e	Electron mass	
m_i	Ion mass	
m_n	Neutral gas atom/molecule mass	
M	Mach number	4.2.1
M_e	Electron magnetization parameter	9.1
M_i	Ion magnetization parameter	9.1
M_d	Dust magnetization parameter	9.1
p	Electric dipole moment	3.3.2
p	Gas pressure	
P	Havnes parameter	2.6
P	Projection matrix	10.6.1
q	Wave vector	
q	Mie scattering parameter	10.7.1
Q_d	Dust charge	2.4.2
r_0	Normalized distance in dust clusters	8.3
R	Rotation matrix	10.6.1
$S(\omega)$	Spectral power density	7.8, 8.6
t	Time	
$\mathbf{t}$	Translation vector	10.6.1
T_d	Dust temperature	5.7
T_e	Electron temperature	
T_i	Ion temperature	
T_n	Neutral gas temperature	
u_i	Ion drift velocity	
$U(x, y)$	Reconstructed hologram	10.5
v_B	Bohm velocity	2.4.3
v_d	Dust velocity	
v_e	Electron velocity	
v_i	Ion velocity	
v_n	Neutral gas velocity	
$v_{th,d}$	Dust thermal velocity	2.2.1
$v_{th,e}$	Electron thermal velocity	2.2.1
$v_{th,i}$	Ion thermal velocity	2.2.1
$v_{th,n}$	Neutral gas thermal velocity	2.2.1
V	Dust interaction potential energy	4.1
Z_d	Dust charge number	2.4.2
Z_f	Charge of ion focus	4.3
α	Electron-to-ion density ratio	2.4.1
β	Epstein friction coefficient	3.4

(continued)

Symbol	Description	See Section
Γ	Coulomb coupling parameter	5.1
δ	Millikan/Epstein reflection parameter	3.4
δ^{ph}	Photoelectron yield	2.3
δ^s	Secondary electron emission yield	2.3
ϵ_0	Vacuum dielectric permittivity	
ϵ	Dielectric function	4.2.1, 6.4
ϵ	Relative dust density	6.1
κ	Screening strength	5.2
λ	Wave length	
λ_D	(Linearized) Debye length	4.1
$\lambda_{D,e}$	Electron Debye length	4.1
$\lambda_{D,i}$	Ion Debye length	4.1
λ_s	Shielding Debye length	7.9
$\ln \Lambda$	Coulomb logarithm	3.3.1
μ	Mach cone opening angle	7.6
μ	Atomic mass number	3.3.2
μ_i	Ion mobility	9.2
ν_{en}	Electron-neutral collision frequency	9.1
ν_{in}	Ion-neutral collision frequency	6.4, 9.1
ϱ_d	Dust mass density	3.1
σ	Absorption cross-section	10.7.2
σ_{in}	Ion-neutral collision cross-section	2.2.4
τ_e	Dust-charging time scale due to electron current	2.5
τ_i	Dust-charging time scale due to ion current	2.5
ϕ_{fl}	Floating potential of dust particle	2.1
ϕ_p	Potential of dust particle	2.2.1
$\hat{\phi}$	Normalized particle floating potential	2.4.1
ψ	Plasma potential	2.6
$\hat{\psi}$	Normalized plasma potential	2.6
ω_0	Confinement frequency	8.3
ω_{res}	Resonance frequency	3.8.1
ω_{cd}	Dust cyclotron frequency	6.5, 9.1
ω_{ce}	Electron cyclotron frequency	9.1
ω_{ci}	Ion cyclotron frequency	6.5, 9.1
ω_{pd}	Dust plasma frequency	6.1
ω_{pe}	Electron plasma frequency	6.1
ω_{pi}	Ion plasma frequency	6.1
Ω	Angular rotation velocity	9.2, 12.4

Index

Abel inversion, 195
Absorption, 193
Acceleration, 51, 175, 220
Accretion, 204
Acetylene, 168, 191, 199
Agglomeration, 203
Alignment, *see* Vertical alignment
Ambipolar diffusion, 35
Ambipolar electric field, 44, 108, 165
Ambipolar ion flow, 45
Amorphous silicon, 209
Angle-resolved scattering, 191
Angular correlation function, 90, 95
Aperture, 173
Ashkin, A., 41

Backward wave, 130
Bandpass filter, 174
Beer-Lambert, 193
Bohm velocity, 21, 51, 135
Boltzmann factor, 13, 27, 103, 110
Bose-Einstein condensate, 41
Bremsstrahlung, 207
B-ring, 220
Brownian motion, 131, 150

Calibration target, 184
Camera, 173
    calibration, 183
    lens, 173
    matrix, 183
    model, 183
Capacitance, 19, 24

Carbon, 207
Cassini spacecraft, 218, 223
Catalysis, 208
Chandrasekhar function, 38
Charge, 48
    gradient, 51, 54
    measurement, 48–55, 61, 71, 124, 125, 127, 151, 155
    number, 19, 20, 105, 204
    position-dependent, 51, 54
Charge density, 102
    dust, 28, 102, 111, 113
    electron, 103
Charged molecules, 201
Charge-to-mass ratio, 47
Charging, 7, 213
    charging time, 23, 24, 26, 54, 221
    current, 8, 217
    delayed, 54, 169
    energy barrier, 13
    fluctuations, 26, 204
    impact parameter, 36
    OML, 8–20
Chip manufacturing, 199
Clustering algorithm, 187
Collision
    charge-exchange, 13, 15
    Coulomb, 36, 67, 78
    dust-dust, 55, 60, 133
    dust-neutral, 35, 162
    electron-neutral, 162
    ion-electron, 36
    ion-neutral, 7, 13, 14, 22, 39, 45, 48, 65, 108, 162, 164, 165, 167
Collisional ion current, 21

© Springer Nature Switzerland AG 2019
A. Melzer, *Physics of Dusty Plasmas*, Lecture Notes in Physics 962,
https://doi.org/10.1007/978-3-030-20260-6

Collisionless, 8, 13, 39, 64
Colloidal plasma, 1
Colloidal suspensions, 1, 139, 174
Complex fluid, 1, 59
Complex plasma, 1, 83
Confinement potential, 50, 60, 61, 86, 115,
        129, 130, 139–142, 144, 148, 168
Continuity equation, 103, 110
Cooper pairing, 64
Coulomb cluster, *see* Dust cluster
Coulomb coupling parameter, 81, 83, 84
Coulomb crystallization, 1, 82, 84, 85
Coulomb logarithm, 36, 38
Cross correlation, 176
Cross section
    absorption, 194
    charge exchange, 13, 15
    Coulomb, 36
    electron collection, 12
    geometric, 10, 40, 41, 172
    ion collection, 10, 21, 36
    scattering, 195
Cyclotron frequency
    dust, 112, 162
    electron, 162
    ion, 112, 162

Damping length, 106, 122, 125
Debye frequency, 117
Debye-Hückel equation, 34
Debye-Hückel potential, 59, 61, 83
Debye length, 13, 34, 37
    electron, 34, 60, 124, 125, 135
    ion, 60, 135
    linearized, 59, 135
Defect, 87, 90, 98
Density
    dust, 27, 85, 102–104, 106, 111, 190, 193,
            200, 204, 207
    electron, 18, 29, 34, 51, 103, 110, 111
    ion, 18, 34, 64, 66, 85, 103, 167
    plasma, 27, 47, 61, 166, 220
Depth of field, 181, 183
Dielectric constant, 33
Dielectric function, 62, 63, 65, 108
Dione, 218
Dipole moment, 33, 62
Discharge
    dc, 106, 108, 164
    radio-frequency, 24, 42, 106, 108, 156, 166,
            200, 208
    silane, 44, 209

Doppler shift, 63, 108
Drift velocity, 14, 20, 34–38, 135, 164
Dust-acoustic wave, 101, 222
    dispersion, 104, 123
    experiments, 106
    friction, 105
    self-excited, 107
    sound speed, 105
Dust cluster, 139, 156
    energy, 142
    magic number, 152, 153
    normal modes, 146–156
    1D, 140
    periodic table, 141, 145
    phase transition, 154
    stability, 152
    2D, 141
    3D, 144
    zigzag transition, 140
Dust-cyclotron wave, 112
Dust-density wave, 108, 221
Dust detector, 216, 218
Dust growth, 2, 26, 44, 168, 191, 199
Dust-in-plasma, 134
Dust ion-acoustic wave, 110
    dispersion relation, 110
    experiments, 111
    sound speed, 111
Dust ion-cyclotron wave, 113
Dust lattice wave, 115
    compressional, 115, 119
    dispersion, 123
    experiments, 122–125
    1D, 115
    1D dispersion, 117
    out-of-plane, 115, 129
    shear, 115, 120, 125
    sound speed, 118, 120, 125
    transverse, 115, 129, 130
    2D, 119, 125
Dust plasma frequency, *see* Plasma frequency,
        dust
Dust removal, 205
Dynamical matrix, 147

E-cross-B drift, 164, 165, 221
Eigen mode, 148, 150
Eigen value, 148
Electron depletion, 28, 85, 190, 204
Electron pressure, 111
Epipolar line, 185
Epstein friction, 40, 49, 76, 78, 108, 162

Equilibrium, 8, 24–26, 54, 103, 130, 146, 148, 156, 217, 220
Equilibrium position, 45, 46, 73, 95, 96, 115, 122, 125, 143, 147, 154

Faraday cup, 55
Filaments, 166
Fine-particle plasma, 1
Floating potential, 8, 17, 18, 28
    multivalued, 22
Force
    attractive, 61, 62, 67, 73, 78, 96
    collection, 36, 38
    Coriolis, 169
    Coulomb, 36, 38, 130
    drag, 35
    electric field, 31, 102, 140, 141, 144, 164, 165, 206, 216, 219
    friction, 39
    gravity, 31, 140, 141, 164, 165, 213, 216, 219
    ion drag, 35, 164, 165, 168
    Langevin, 96
    Lorentz, 161, 169, 216
    neutral drag, 39, 164, 165, 206
    non-reciprocal, 67, 69, 73, 78, 96
    photophoretic, 42
    polarization, 33
    radiation pressure, 41, 69, 156
    repulsive, 69
    thermophoresis, 40, 144, 206
Frame rate, 175
Fresnel-Kirchhoff integral, 179
Fresnel rings, 179
Fundamental matrix, 186
Fusion, 206

Galileo spacecraft, 216
Gravitational acceleration, 31
Gravito-electrodynamics, 213
Gyro-orbit, 162

Hall parameter, 162
Halo orbits, 219
Harmonic expansion, 147
Harmonic oscillator, 47, 75
Harmonic potential, 51
Havnes effect, 28, 85, 190, 204
Havnes parameter, 28
Heating, 41, 45, 95, 96, 154, 156, 206
Hexagonal, 73, 83, 86, 90, 96, 119, 121, 152

Hexatic phase, 88
HMDSO, 199
Hologram, 179
Holography, 179
Huygens principle, 179

Image analysis, 173
Image charges, 204
Impact parameter, 9
In-line holography, 179
Integrated circuits, 205
Interaction potential, 60
Interference filter, 171
Io, 216
Ion focus, 67–69, 71, 72, 77, 95, 154, 167, 168
Ion inertia, 111
Ion shadow, 167

Jupiter, 216

Kalman filter, 175, 187
Kepler, 15, 220, 222
Killer particle, 2, 205
Krakatoa, 214

Landau damping, 65
Lane formation, 190
Langmuir, 8
Langmuir probe, 7
Larmor radius, 161
Laser heating, 156
Laser manipulation, 41, 49, 69, 122, 125, 128
Levitation, 32, 41, 42, 144, 220
LIDAR, 215
Light-field camera, 177
Linear chain, 73, 115, 122, 130
Line of sight, 184

Mach cone, 127, 222
    compressional, 129
    secondary, 129
    shear, 129
Mach number, 62
Madelung energy, 83
Magic number, 139, 152, 153
Magnetization parameter, 162
Mass density, 31, 111
Mass spectrometry, 201
Mathieu's equation, 51

Maxwellian
    distribution, 10, 12–14, 16, 17
    shifted, 38
Mean free path, 15, 22, 66
Melting transition, 79, 83, 90, 94–96, 155–158
Methane, 199
MF particles, 45, 48
Microgravity, 31, 32, 43, 44, 91, 107, 129
Mie scattering, 171, 190, 192, 194, 195
Millikan's experiment, 40
Mobility, 164
Mode
    breathing, 148
    intershell rotation, 153
    rotation, 148
    sloshing, 148
    spectrum, 151, 158
    vortex-antivortex, 153
Mode-coupling instability, 72, 95
Modulation depth, 51
Molecular cluster, 200
Moment method, 174
Moment method with bandpass, 174
Momentum equation, 103, 110
Momentum transfer, 35
Multiple scattering, 190
Multiset triangulation, 181, 184, 188

Newton's third law, 67, 70
Noctilucent clouds, 213
Non-reciprocal attraction, 67, 69, 96
Nucleation, 201

Object wave, 179
OCP, 81
OML, 8, 21, 22, 162
OML current, 8, 11, 17, 26
One-component plasma, *see* OCP
Onion shell structure, 144
Open system, 67, 78
Optical trap, 42
Optical wave, 131
Orbital motion limit, *see* OML

Pair correlation function, 87, 95
Parabolic flight, 43, 181, 184, 189
Particle correspondence, 187
Particle image velocimetry, 176
Particle-particle distance, 4, 65, 66, 82, 85, 88,
    91, 115, 121, 124, 133, 141, 143,
    166, 171, 178

Particle tracking, 122
Permutation, 187
Phase transition, 94–98, 164
    defects, 98
    dust cluster, 154
Phonon, 132
Photoemission, 16, 55, 218, 220
PIV, 176
Pixel, 173, 174, 185, 191, 192
Pixel locking, 173
Plasma crystal, 3, 61, 65, 69, 72, 73, 78, 79,
    84–86, 106, 118, 125, 127, 129
Plasma facing component, 206
Plasma frequency, 162
    dust, 74, 78, 102, 105, 109, 117
    ion, 101, 110, 111
Plasma potential, 8, 27, 28, 44, 47, 201
Plenoptic camera, 177
Point charge, 69, 71, 82, 96
Poisson's equation, 47, 103, 104, 110
Polarization degree, 190
Polarization state, 171
Polar mesospheric summer echoes, 213
Polymerization, 199
Polymorphous silicon, 3, 210
Potential well, 15, 47, 48, 51–53, 86, 131
Projection matrix, 183
Pseudo-magnetization, 169

Quasineutrality, 25, 28, 32, 47, 103, 110, 111,
    113, 221

Radar, 215
Radial motion theory, 13
Rayleigh scattering, 172
Reconstruction plane, 180
Reference wave, 179
Residue image, 188
Resonance, 47, 49, 51, 53, 55, 71, 124, 125,
    131, 142, 168
    phase-resolved, 49
Rhea, 218

Saturn, 1, 218
Savitzky-Golay filter, 175
Scaling parameter, 143, 151
Scanning mirror, 178
Scanning video microscopy, 178
Scattering, 89
Scattering parameter, 190
Schweigert instability, 154
Schweigert model, 69, 71

Screening strength, 83, 105, 117, 124, 143, 148, 155
Secondary electron emission, 16, 22
Shake-the-box algorithm, 181, 188
Sheath, 7, 13, 18, 31, 32, 34, 43, 45–47, 49, 51, 54, 62, 69, 86, 92, 108, 115, 133, 135, 140, 154, 164–166, 205, 208
  radio-frequency, 25, 47, 65
Sheath electric field, 108
Shell occupation, 146
Shielding cloud, 32
Shielding length, 34, 39, 133, 135
Showerhead electrode, 206
Silane, 44, 199, 209
Silicon, 199, 208
Silicon clusters, 203
Silyl, 202
Silylene, 202
Solar cell, 209
Sound speed, 127, 222
Sound wave, 111
Spectral power density, 133, 150, 155, 158
Spitzer, 18
Spokes, 2, 219
Spring constant, 71, 115, 117, 147
Sputtering, 205, 208
Stereo matching, 185
Stereoscopy, 181
Strong coupling, 3, 4, 82, 85, 88, 117, 222
Structure factor, 90
Subpixel resolution, 173
Superconducting magnet, 162
Supersonic flow, 34, 62, 67, 70, 135
Supersonic speed, 35, 127, 128, 222

Temperature
  atmosphere, 214
  dust, 95, 96, 98, 156
  dust surface, 93
  electron, 7, 9, 20, 28, 48, 95, 135, 208
  gradient, 41, 43, 144
  ion, 9, 10, 105, 135
  mode, 150
  neutral, 40
  photoelectron, 16
  secondary electron, 17
Thermal velocity
  dust, 104
  ion, 11, 14, 21, 36, 38, 136
  neutral, 40

Thin-film deposition, 204, 205
Threshold
  friction, 76, 78
  modulation, 51
  pixel intensity, 174, 189
Trajectory linking, 187
Trap, 1, 3, 13, 15, 24, 32, 41, 42, 45, 47, 48, 59, 62, 69, 86, 106, 115, 127, 139, 142, 156, 163, 166, 205

UV radiation, 55

Vertical alignment, 61, 62, 69, 75, 77, 79, 154
  instability of, 72–79
  oscillatory instability, 76, 78
Vertical resonance, 69
  linear, 47, 48
  nonlinear, 51
  parametric, 50
  self-excited, 54
Void, 35, 45, 144, 191

Wakefield, 62, 65, 67, 68, 95, 154, 167, 168
  magnetized, 167
Water ice, 214
Wave damping, 109, 118
Waves
  dust-acoustic (see Dust-acoustic wave)
  dust-cyclotron (see Dust-cyclotron wave)
  dust ion-acoustic (see Dust ion-acoustic wave)
  dust ion-cyclotron (see Dust ion-cyclotron wave)
  dust lattice (see Dust lattice wave)
Wiener filter, 175
Wigner–Seitz cell, 87, 93, 144
Wigner-Seitz radius, 82, 83
Wire excitation, 50, 53, 60, 106
Wire mesh, 166

Yukawa ball, 41, 144, 156
Yukawa cluster, see Dust cluster
Yukawa potential, 59, 61, 83, 119, 142, 146
Yukawa system, 83, 91

Printed in the United States
By Bookmasters